BK 621.381   Z46T
TRANSISTOR CIRCUIT ANALYSIS AND APPLICATION
                              /ZEINES, BE
    C1976   16.95 FV

3000 397612 30017
St. Louis Community College

CO-ASV-776

WITHDRAWN

FV

621.381 Z46t
ZEINES
  TRANSISTOR CIRCUIT ANALYSIS
  AND APPLICATION
                        16.95

St. Louis Community
College

Library

5801 Wilson Avenue
St. Louis, Missouri 63110

TRANSISTOR CIRCUIT ANALYSIS AND APPLICATION

TRANSISTOR CIRCUIT ANALYSIS AND APPLICATION

BEN ZEINES

RESTON PUBLISHING COMPANY, INC., Reston, Virginia

A Prentice-Hall Company

Library of Congress Cataloging in Publication Data

Zeines, Benjamin.
    Transistor circuit analysis and application.

    Includes index.
    1. Transistor circuits.   I. Title.
TK7871.9.Z44   621.3815'3'0422   75-35697
ISBN 0-87909-837-6

©1976 by
RESTON PUBLISHING COMPANY, INC.
A Prentice-Hall Company
Reston, Virginia

All rights reserved. No part of this book
may be reproduced in any way, or by any
means, without permission in writing from
the publisher.

10   9   8   7   6   5   4   3   2   1

Printed in the United States of America.

PREFACE

The purpose of this text is to provide a comprehensive treatment on transistor circuits and applications. This text is suitable for technical institutes, community college engineering colleges, industrial and military training programs and engineering retraining programs. The mathematical prerequisites for all students are college algebra, trigonometry, basic ac and dc circuit theory and the ability to construct a straight line.

The material is presented in a straightforward and logical manner using a step-by-step procedure. The first two chapters deal with the development of semiconductor physics and the use of the semiconductor diode and transistor as a circuit element. Chapter 3 deals with the various equivalent circuit and mathematical models representing the transistor. The advantages and disadvantages of each are discussed. Chapter 4 introduces the usage of transistors in audio cirucits, wide band amplifier circuits and tuned voltage amplifier circuits. Equal coverage is given to the bipolar junction transistor and the field effect transistor. A troubleshooting analysis follows each topic discussion pointing out possible troubles after circuit construction and possible solutions to the problem.

The following chapters discuss power supplies, power amplifiers, feedback amplifiers and oscillators, modulation and demodulation theory and analysis.

In all cases, practical circuits used in field equipment are specified to illustrate the theory and principles involved. There are numerous problems at the end of almost every chapter as well as many illustrative problems interwoven with the theory and analysis section. In spite of all this, the emphasis is qualitative rather than quantitative since the problems and sample problems are designed to demonstrate the application of theory in analyzing and understanding networks and amplifier performance.

Many texts have been written for the field of semiconductors but one more is presented here for a specific reason. There is a gap between texts written for high school students and those written for the engineering university student. It is hoped that this text will bridge the gap.

Ben Zeines

CONTENTS

three   **SMALL SIGNAL AMPLIFIERS**   75

four   **TRANSISTORIZED VOLTAGE AMPLIFIERS**   95

one

SEMICONDUCTOR DIODES

INTRODUCTION

The advent of semiconductors in 1948 expanded the science of electrical technology many fold. In years prior to semiconductors, the processes of electronic signal amplification, generation, waveshaping and switching were generally performed by the vacuum tube. Since 1948, the semiconductor has proven it can perform these same functions with greater efficiency and consequently has replaced the vacuum tube in many fields of endeavor.

Some of the basic characteristics of semiconductors are:

1. Small physical size
2. Mechanically rugged

3. Extremely long life

4. Require no filamentary power

The study of semiconductor technology requires the understanding of the basic physics of the structure of matter.

ATOMIC THEORY

The current theory is that matter is composed of tiny particles called *atoms*. An *element* is a substance wherein all the atoms contained are of the *same kind*. Substances containing different kinds of atoms are called *compounds*. It should be noted that a compound can be separated into its component elements, and an element cannot be further separated.

The atom can also be subdivided into tinier particles and resembles the structure of the earth's solar system. An atom contains a nucleus comprising positive charges called protons and a number of neutrons having zero charge. Surrounding the nucleus are a number of negatively charged particles called *electrons*. The accepted theory is that the electrons rotate about the nucleus in elliptical orbits similar to the motion of the planets about the sun. A typical carbon atom is shown in Fig. 1-1.

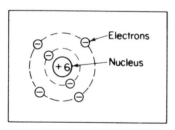

Figure 1-1 *Carbon Atom*

The mass of the atom is contained almost wholly in the nucleus. The mass of the proton is approximately 1840 times the mass of the electron. On the other hand, the electron charge is 1.6×10^{-19} coulombs and is equal and opposite to the charge of one proton.

The attractive force between electron and proton is counterbalanced and a discrete distance maintained between them because of the outward centrifugal force developed by the speed of the orbiting electron. This is the reason the electrons are generally not devoured by the nucleus nor do they fly into outer space.

Each element is different in that the number of electrons and protons are uniquely associated with that element only.

The *atomic number* of an element gives the number of protons within the nucleus. Every atom contains one or more electrons orbiting around the nucleus.

The physicist assumes the orbiting path of the electron is three dimensional and, consequently, the term *shell* is used to describe the possible paths of orbit.

The shells are finite in character and the electrons of a specific element must exist within one of the given shells. No electrons lie or exist in between shells. To cause an electron from one shell to shift to another shell, discrete amounts of energy called *quanta* must be applied. A *quantum of energy* is the smallest amount of energy required for the process. Quanta of energy must exist as whole integral numbers since fractional parts of quantum do not exist.

The outer shell called the *valence shell* determines the chemical activity of the element. If the outer shell is completely filled, the substance is considered chemically inert and does not enter readily into chemical reactions. On the other hand, an incomplete outer shell in an element may result in chemical combinations to produce the effect of a filled outer shell. This action produces molecules of stable chemical compounds such as salt, water and so forth.

The capacity for electrons in each shell is in accordance with the formula $2n^2$, where n is the number of the shell counting from zero (which is the nucleus). Thus, the maximum capacity of the first shell is 2, the second shell is 8, the third shell is 18 and so on. Table 1-1 illustrates the theory of electron arrangement in shells.

TABLE 1-1   *ELECTRON SHELL ARRANGEMENT IN ATOMS*

Shell	Maximum number of electrons
1	2 electrons
2	8 electrons
3	18 electrons
4	32 electrons
5	50 electrons

The electron distribution in the shell layers can be determined from two basic rules.

Rule 1:   The maximum number of electrons in the outermost shell layer of any element cannot exceed eight.

Rule 2:   The maximum number of electrons in the shell prior to the outermost shell layer can never exceed 18.

Knowledge of the atomic number will permit the arrangement of the electrons into the proper shell structure. An illustrative problem will demonstrate the theory.

sample problem

The atomic number of copper is 29. Determine the number of electrons per shell layer.

Solution:

The nucleus has 29 protons and 29 orbiting electrons. These electrons follow the sequence of:

first layer— 2 electrons
second layer— 8 electrons
third layer—18 electrons
fourth layer— 1 electron
29 electrons

The atomic structure of the copper atom is:

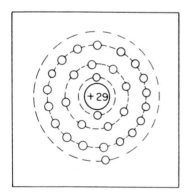

sample problem

Determine the atomic structure of silver with an atomic number of 47.

Solution:

first layer— 2
second layer— 8
third layer—18
fourth layer—18
fifth layer— 1

ELECTRON ENERGY BONDS

Certain elements have the property of assuming a stable crystalline form. This internal regularity results in the element displaying straight sides or faces. In the crystalline form (called a crystal lattice) valence electrons of two atoms interlock or share electrons. Germanium and silicon have the ability to form covalent bonds. Figure 1-2 illustrates the structure of the germanium crystal.

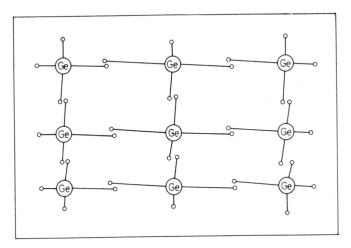

Figure 1-2 *Covalent Bonds in Germanium*

The lines from the electron to the nucleus represent the attraction of the nucleus on its electrons. Note that each electron orbits in such a manner that it is shared by two atoms.

ELECTRON ENERGY LEVELS

The position of the electrons with respect to the nucleus determines the energy level of these electrons. If the position of the electron is far from the nucleus, a small amount of energy will be adequate to set it free. As the position of the electron is moved closer to the nucleus, a vast amount of energy is necessary to free the electron from the nucleus. The energy of the electron is specified in terms of electron volts. An electron volt is the amount of energy acquired by an electron when it falls through a difference of potential of one volt.

In order to have current flow in a material the electron must escape from its atomic bond and move into the so-called conduction region. The application of an external emf creates an electric field providing energy to the atom causing an electron to move out of its valence band into the conduction region. This motion results in a current flow.

In general, all elements may be categorized into three major classifications: 1. *conductors*, 2. *semiconductors* and 3. *insulators*. Energy diagrams are usually used to illustrate the basic three categories and are shown in Fig. 1-3.

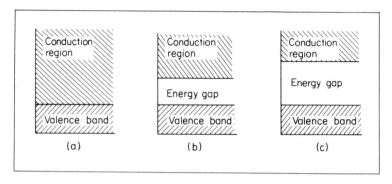

Figure 1-3   *Energy Diagrams: (a) Conductor (b) Semiconductor (c) Insulator*

In the energy diagram labeled conductor, it is apparent that the valence band and conduction region almost overlap. The application of a little energy (about .001 ev) is required to cause an electron to escape from the valence band into the conduction region.

The semiconductor energy diagram shows an energy gap existing between the valence band and the conduction region. Consequently, considerable energy (about .7 ev for germanium and 1.1 ev for silicon) must be supplied to the electron in the valence band to cause an electron to leap across the energy gap into the conduction region.

For the insulator, the energy gap is still wider, requiring higher energies (about 7 ev) to cause an electron to jump across the energy gap into the conduction region.

SEMICONDUCTORS

Two semiconductors often used for transistor operation are germanium and silicon. Pure germanium or pure silicon has an intrinsic characteristic: the covalent bonds must be broken before conduction can take place. Energy may be supplied through thermal or light energy.

Whenever an electron acquires sufficient energy to break away from a covalent bond, it leaves behind a vacancy called a "hole." The vacancy or hole exhibits the property of a mobile positive charge. The motion of the hole is opposite to that of the electron. Consequently, in pure germanium or silicon the freeing of an electron is always accompanied by the creation of a hole. Thus, the application of thermal energy to the semiconductor produces electron-hole pairs. The conduction that exists in a pure semiconductor is then called *intrinsic conduction*.

EXTRINSIC CONDUCTION

For transistor operation, however, it is necessary to control the electrical properties of the semiconductor material. "Doping" is the process of adding certain impurities to the semiconductor material. The ratio of impurity to germanium is one atom of impurity to ten million atoms of germanium. The impurity used can be aluminum, indium, gallium, arsenic or phosphorous.

Depending on the type of impurity used, two types of semiconductors will result: N type or P type. The atoms of the impurity element substitute themselves in the crystal lattice for the atoms of the basic semiconductor material. These impurity atoms are called either donor impurities or acceptor impurities.

N TYPE GERMANIUM

Impurities having five valence electrons (such as arsenic or antimony) may be added to germanium in controlled amounts. The result of adding this impurity element to germanium is illustrated in Fig. 1-4.

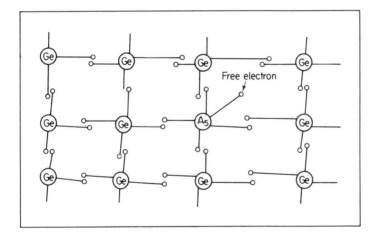

Figure 1-4   *Crystalline Structure of Germanium with Arsenic Impurity*

The ionization energy required to free excess electrons is small compared to the energy levels necessary to cause conduction in a pure semiconductor. Each pentavalent impurity atom added produces an excess electron that is available for conduction.

In addition to the donor electrons, there are also free electrons due to electron-hole pairs in the conduction region and holes from those pairs in the valence band. Consequently, the electrons are termed the majority current carriers and the holes are the minority current carriers in N type germanium.

The pentavalent impurity atom adds or donates free electrons to the semiconductor material and therefore is called a *donor impurity*. The semiconductor material doped in this way is said to be *N type* semiconductor and is symbolized by the capital letter N.

P TYPE GERMANIUM

The addition of trivalent impurity atoms into the semiconductor material is shown in Fig. 1-5.

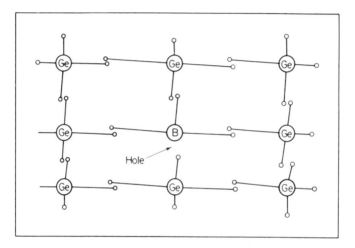

Figure 1-5   *Crystalline Structure of Germanium with a Boron Impurity*

The impurity atom contains three valence electrons and forms a covalent bond with the germanium atom. Since one valence electron is missing, the position normally filled by an electron is designated as a *hole*. The hole has a positive charge and steals an electron from an adjacent atom creating a new hole while filling the old one. Holes, in motion, constitute an electric current.

Impurities that create a hole in germanium or silicon are called *acceptor impurities* because they accept electrons from the crystal. This type of semiconductor is said to be *P type* semiconductor and is symbolized by the letter P.

CONDUCTION IN SEMICONDUCTORS

In Fig. 1-6, a battery is placed across a slab of N type germanium. In N type germanium, the majority carriers are electrons. The direction of the electron current flow is always determined by the external battery. Electron current flows from the negative terminal of the source through the semiconductor to the positive terminal of the source.

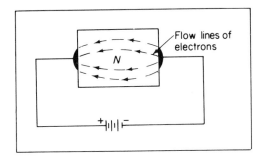

Figure 1-6   *Battery Across N Type Germanium*

In Fig. 1-7, a battery is placed across a slab of P type germanium. In P type material, the majority carriers are holes. The holes are considered positive charges and are attracted to the negative terminal. An external current flows because holes are continuously being created at the positive terminal. The flow lines of holes are also illustrated in Fig. 1-7.

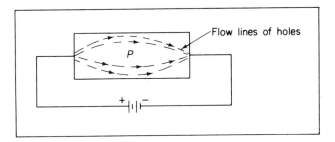

Figure 1-7   *Battery Across P Type Germanium*

PN JUNCTION

When N type germanium is united with P type germanium the contact surface is referred to as a *PN junction*. Some of the holes in the P region and some of the electrons in the N region migrate toward each other and combine. Each combination eliminates an excess electron and hole. This action exists for a short period of time in the immediate vicinity of the junction. The fixed impurity atoms in the P region exhibit a negative charge and repel the free electrons. The fixed impurity atoms in the N region exhibit a positive charge and repel the holes. The potential that exists at the junction because of the unlike charges is commonly called the *potential gradient*, or potential energy barrier, as shown in Fig. 1-8.

The PN junction contains the space charge battery and this region is also referred to as the *depletion region*. The physical distance of the space

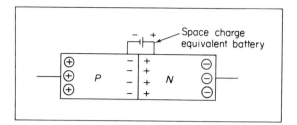

Figure 1-8 *PN Junction*

charge region is called the width of the depletion region and is in the order of one micron. Figure 1-9 illustrates the PN junction with the depletion region.

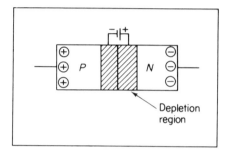

Figure 1-9 *Depletion Region in PN Junction*

BIAS CONSIDERATIONS

An external battery is connected to the PN junction as illustrated in Fig. 1-10.

The holes will move to the negative terminal of the battery E. Simul-

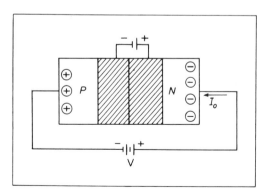

Figure 1-10 *Reverse Biased PN Junction*

taneously, the electrons will move to the positive terminal of the battery. This motion of holes and electrons tends to widen the depletion region until the potential of the space charge battery is equal to the potential of the external battery. Consequently, in the ideal case, no current flows and the PN junction is said to be reverse biased. In the practical case, a small current is produced by the minority carrier flow and is defined by the terms leakage current, reverse bias current, saturation current, or cutoff current. The symbol for this current is I_0.

Reverse the external battery connections to the PN junction as shown in Fig. 1-11.

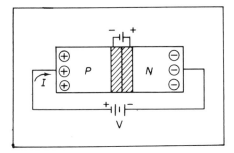

Figure 1-11   *Forward Biased PN Junction*

The electrons present in the N region are repelled by the negative charge applied at the negative terminal of the battery. If the applied voltage is greater than the space charge equivalent battery, some electrons cross the junction and enter the P region. Similarly, holes in the P region cross into the N region where they combine with existent electrons.

For each combination of an excess electron and a hole that occurs, an electron from the external battery enters the N type region and moves toward the junction. Similarly, at the positive terminal, an electron from a covalent pair is set free and moves toward the positive terminal. Consequently, a hole is created that moves toward the junction. The total current flowing through the semiconductor material is composed of electron flow in the N region and hole flow in the P region. The total current flows from the positive terminal of the battery to the negative terminal passing through the PN junction. Connecting the battery in this manner forward biases the PN junction semiconductor.

DIODE ACTION

To utilize the PN junction as a circuit element, a description of its static or dc characteristics is needed. Figure 1-12 is a plot of current flow versus applied voltage of a practical PN junction semiconductor.

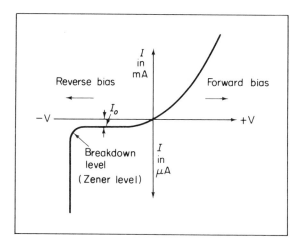

Figure 1-12   *Graph of I Versus Voltage in a PN Junction*

Note that the current flow in the forward bias direction is extremely high since it is measured in mA. The current flow in the reverse bias direction is quite low since it is measured in μA.

At zero applied voltage, the resultant current through the diode is zero. With the application of a small positive voltage the depletion region narrows and current begins to flow. A greater increase in the forward bias results in a larger flow of current. In this region the PN junction acts like a low resistance.

In the reverse bias condition, current rapidly approaches the leakage level (I_0) and remains there. Increasing the negative voltage applied does not yield a greater current until a specific voltage value called the breakdown or zener voltage is applied. At this value, there is a sharp increase in reverse current and the PN junction is said to have *broken down*. The zener voltage is considered to be the maximum reverse bias or maximum peak inverse voltage that the PN junction can withstand.

It is evident that the PN junction is capable of rectification. If an ac signal wave were applied across the PN junction, current would flow during the positive half cycle and there would be little or zero current during the negative half cycle. Therefore, the junction formed by the combination of the P and N type semiconductors acts as an efficient rectifying device and is usually referred to as a *junction diode.*

The symbol for the semiconductor is shown in Fig. 1-13.

The empirical relationship that relates the current to voltage in a semiconductor diode is given by

$$I = I_0(\epsilon^{E/E_T} - 1) \text{ amps.}$$

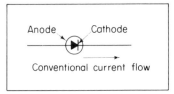

Figure 1-13 *Semiconductor Diode Symbol*

where:

$I_0 =$ leakage current

$E =$ applied voltage in volts

$E_T = \dfrac{kT}{q} = \dfrac{T}{11,600}$ volts

$T =$ absolute temperature in °K.

Positive values of E correspond to forward bias and result in a positive value of I. Negative values of E correspond to reverse bias and give a negative value of I. The diode equation is valid for all values of E up to the zener voltage level.

An illustrative problem will demonstrate the theory.

sample problem

The total leakage current for a PN junction diode is 10 μA at 20°C. Determine the current in the circuit when the applied voltage is 0.2 volt.

Solution:

Step 1: Calculate E_T

$$E_T = \frac{T}{11,600}$$

$$E_T = \frac{293}{11,600}$$

$$E_T = 25.3 \text{ mV}$$

Step 2: Calculate I

$$I = I_0(\epsilon^{E/E_T} - 1)$$

$$\frac{E}{E_T} = 7.9$$

$$\epsilon^{7.9} = 2,700$$

$$I = 10 \times 10^{-6} (2,700 - 1)$$

$$I = 27 \text{ mA}$$

The leakage current of the semiconductor PN junction diode is extremely sensitive to changes in temperature. The variation of leakage current with respect to temperature is shown in Fig. 1-14.

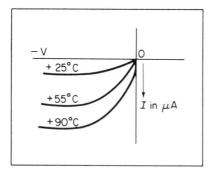

Figure 1-14   *Variation of Leakage Current Versus Temperature*

The mathematical expression for the leakage current as a function of temperature is:

$$I_{0_f} = I_{0_i}\epsilon^{k(T_f - T_i)}$$

where:

I_{0_i} = value of leakage current at the initial temperature
T_i = initial temperature in °C
T_f = final temperature in °C
k = constant as a function of the material (germanium has a value of 0.075, silicon has a value of 0.13)

If the temperature increases by 10°C over the initial temperature in a germanium semiconductor the ratio of I_{0_f} to I_{0_i} is given by

$$\frac{I_{0_f}}{I_{0_i}} = \epsilon^{.75} \cong 2$$

Consequently, as a rough approximation, near room temperature, the leakage current doubles for each increase of 10°C in germanium, whereas for a silicon diode, a 6°C increase will double the leakage current.

An illustrative problem will demonstrate the theory.

sample problem

A silicon diode has a leakage current of 10 μA at 20°C. Determine the change in temperature required for the leakage current to rise to 35 μA.

Solution:

$$I_{0_f} = I_{0_i}\epsilon^{k(\Delta T)}$$

$$35 \times 10^{-6} = 10 \times 10^{-6}\,\epsilon^{.13(\Delta T)}$$

$$3.5 = \epsilon^{.13(\Delta T)}$$

$$\log 3.5 = .13\,(\Delta T)\log\epsilon$$

$$.544 = .13(.434)(\Delta T)$$

$$\Delta T = 9.65°C$$

The dynamic resistance of the semiconductor PN junction diode is usually defined for the forward biased region as equal to:

$$r = \frac{\Delta E}{\Delta I}$$

$$r = \frac{E_T}{I_0\epsilon^{E/E_T}} = \frac{E_T}{I + I_0}$$

An illustrative problem will demonstrate the theory.

sample problem

The total leakage current for a PN junction diode is 10 μA at 20°C. The applied voltage is 0.2 volt. Determine the dynamic resistance.

Solution:

Step 1: Calculate E_T

$$E_T = \frac{T}{11,600}$$

$$E_T = \frac{293}{11,600}$$

$$E_T = 25.3 \text{ mV}$$

Step 2: Calculate I

$$I = I\,(\epsilon^{E/E_T} - 1)$$

$$I = 10 \times 10^{-6}(\epsilon^{7.9} - 1)$$

$$I = 27 \text{ mA}$$

Step 3: Calculate r

$$r = \frac{E_T}{I + I_0}$$

$$r = \frac{25.3 \text{ mV}}{27 \text{ mA}}$$

$$r = .935 \ \Omega$$

JUNCTION DIODE LOAD LINE

The basic diode circuit of Fig. 1-14a consists of a PN junction diode in series with a load resistance R_L and an input signal source v_i.

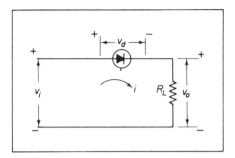

Figure 1-14a *The Basic Diode Circuit*

The circuit is now analyzed to determine the instantaneous diode current and voltage when the input voltage is applied. Using Kirchhoff's voltage law,

$$v_d = v_i - i R_L$$

Consequently, this equation contains two unknowns, namely: v_d and i. A second relation existent between these two variables is given by the static characteristics of the diode as shown in Fig. 1-14b.

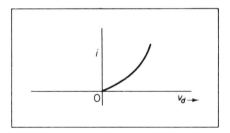

Figure 1-14b *Static Characteristics of a PN Junction Diode*

The expression for the diode voltage is the equation of a straight line called the *load line*. The construction of the load line is performed by means of the slope intercept method. Thus,

$$\text{Let} \quad i = 0 \qquad v_i = v_d$$

$$\text{Let} \quad v_d = 0 \qquad i = \frac{v_i}{R_L}$$

Place these values on the static characteristics and connect the two points together as shown in Fig. 1-14c.

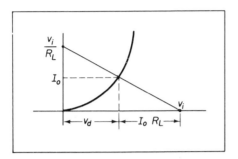

Figure 1-14c *Load Line Construction*

The point of intersection between the load line and the static characteristics defines the point of operation. This denotes the current that will flow in the circuit under these load conditions.

VARACTORS

The characteristics of a junction diode are modified in the high frequency region by the presence of a capacitor. Forward biasing the PN semiconductor diode decreases the width of the depletion region. This results in a relatively high capacitance across the depletion region, called the diffusion or storage capacitance. The equivalent circuit of the forward biased PN junction diode is shown in Fig. 1-15.

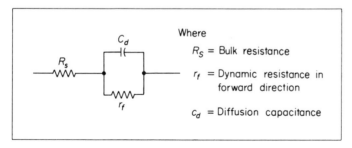

Figure 1-15 *High Frequency Equivalent Circuit of Forward Biased PN Junction Diode*

If a reverse bias is applied to the PN junction diode the voltage appears essentially across the depletion region. Consequently, the depletion region acts as a capacitor separated by a dielectric. Increasing the reverse bias widens the depletion region and decreases the capacitance. This capacitance that exists across the depletion region is called the space charge or transition capacitance.

This space charge capacitance does not affect circuit operation at low frequencies. However, at high frequencies the capacitance acts as a circuit element. The equivalent circuit of the diode in the reverse bias region is shown in Fig. 1-16.

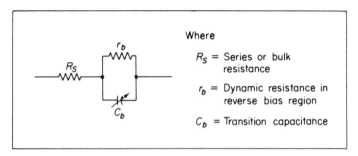

Figure 1-16   *High Frequency Equivalent Circuit of Reverse Biased PN Junction Diode*

It is evident from the diode analysis that the transition capacitance varies inversely with the applied bias voltage. Diodes designed for usage as voltage variable capacitors are called *varactors, varicaps* or *voltacaps* and are symbolized as shown in Fig. 1-17.

Figure 1-17   *Circuit Symbol for the Varactor Diode*

A possible circuit application of the varactor is shown in Fig. 1-18.

The diode is reverse biased by the reduction of V_{dd} through R. An *RF* choke applies the bias to the diode and maintains a high impedance at the frequency of the Hartley oscillator. The capacitance of the varactor diode tunes the tank circuit formed by L_1 and L_2. Increasing the bias lowers the tuning capacitor and increases the frequency of oscillation. A minimum value of reverse bias is necessary to prevent the positive portions of the oscillator signal to forward bias the varactor.

ZENER DIODES

Examination of the circuit characteristics of a PN junction diode shown in Fig. 1-19 indicates that the voltage drop across the diode is constant once the breakdown point is reached. Two physical possibilities are recognized as

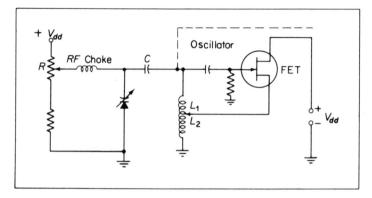

Figure 1-18 *Application of Varactor Diode*

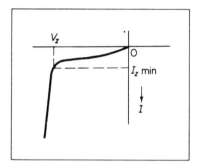

Figure 1-19 *Reverse Biased PN Junction Diode Characteristic*

responsible for the PN junction diode breakdown: *avalanche breakdown* and tunneling or *zener level* effects. Avalanche breakdown occurs in PN junction diodes having breakdown voltages in excess of five volts. Below five volts tunneling effects predominate.

AVALANCHE BREAKDOWN

Consider the leakage current flowing across the depletion region of the PN junction diode. Current carriers comprising the leakage current create electron hole pairs. As the reverse voltage is increased, this generation process accelerates and may produce additional electron hole pairs. This process is cumulative and is called the *multiplication process* or *avalanche multiplication*. As the reverse bias is increased above the breakdown voltage point, the leakage current increases rapidly as shown in Fig. 1-19.

ZENER BREAKDOWN

Zener breakdown occurs in PN junction diodes having a large number of impurities in both the N and P regions. When the applied reverse bias reaches a specified level, the covalent bonds are broken or an electron escapes from a covalent bond creating an electron hole pair. The free electrons move from the P side to the conduction N region. This process may be considered as a *voltage assisted tunneling* since it requires an external electric field.

The term zener is commonly used to describe both PN junction diodes operating in the avalanche or breakdown modes as well as the Zener region. The symbol and equivalent circuit for the zener diode is shown in Fig. 1-20.

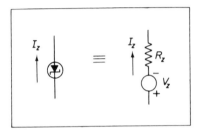

Figure 1-20 *Symbol and Equivalent Circuit for Zener Diodes*

It should be noted that once the Zener level is exceeded, the voltage drop across the diode is relatively constant. The ability of the device to maintain a constant voltage is used in a voltage regulator circuit as shown in Fig. 1-21.

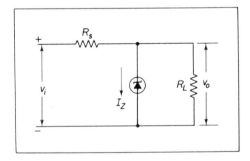

Figure 1-21 *Shunt Connected Zener Diode*

The difference in voltage between the input V_i and the output V_0 is dropped across R_s. The diode will now maintain a constant output voltage despite input voltage changes or load current variations. As the input voltage or load current varies, the diode current will change accordingly to maintain a relatively constant output. The limits of diode current regulation are $I_{z_{min}}$ specified by the knee of the curve and $I_{z_{max}}$ which is determined by the power handling capability of the diode.

TUNNEL DIODE

The tunnel or Esaki diode is a highly doped junction diode that exhibits a negative resistance characteristic over part of its operating range as shown in Fig. 1-22.

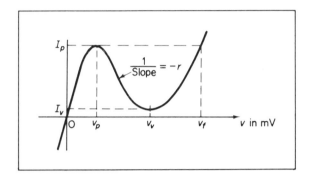

Figure 1-22   *Volt Ampere Characteristic of a Tunnel Diode*

Note that a small forward bias causes current to rise to a peak value corresponding to I_p. As voltage is increased from V_p, the current decreases to a minimum value corresponding to I_v. The tunnel diode, therefore, exhibits a negative resistance characteristic between the *peak current*, I_p, and the *valley current*, I_v. As voltage is further increased past V_v, the current reaches the peak current value I_p at a voltage called the *peak forward voltage* V_f. Further increase of voltage past V_f yields the typical diode characteristic.

A negative resistance device supplies power rather than consumes it and therefore can be utilized as an amplifier or signal generator.

theory

An energy level diagram representative of the N and P type semiconductor is shown in Fig. 1-23.

When the P and N material are joined, the energy level diagram becomes the one shown in Fig. 1-24. The junction barrier produces an alignment of the N and P type materials. If the doping of the N and P type materials is very heavy, the depletion region narrows and the potential hill increases as shown in Fig. 1-25.

Note that as a result of moving the N region downward, the P region valence band and the N conduction region are precisely aligned. These conditions produce a transfer of electrons through the junction by the phenomenon called *tunneling*. The electrons that tunnel through the junction operate with no loss in energy. With zero bias applied, a condition of equilibrium exists in which the total number of electrons entering the P region is exactly equal to the number of electrons leaving the P region. As

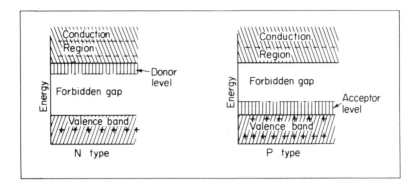

Figure 1-23   *Energy Bands for the N and P Type Material*

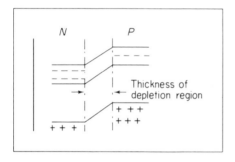

Figure 1-24   *Energy Diagram of PN Junction Diode*

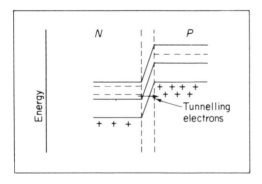

Figure 1-25   *Energy Diagram for a Forward Biased PN Junction Diode*

forward bias is increased from zero, the tunnel current increases and the diode behaves as a low-value resistor. In addition, the forward bias shifts the N region upward relative to the P region. When this occurs the conduction band is aligned with a forbidden gap region with a consequent reduction in current flow.

The tunnel diode current reaches a peak, but then reduces to a minimum or valley condition as the conduction electrons lie opposite a forbidden gap region.

As greater forward bias is applied the conduction band electrons are permitted to cross over the junction barrier and forward bias current rises in the conventional manner. Tunnel diode characteristics are relatively insensitive to both the temperature variations and atomic or nuclear radiation.

The negative resistance characteristic of tunnel diodes makes them useful in amplifier circuits, converters, oscillators and switching circuits. The standard circuit symbol for the tunnel diode is shown in Fig. 1-26.

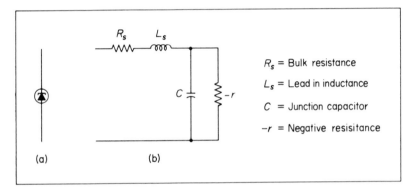

R_s = Bulk resistance

L_s = Lead in inductance

C = Junction capacitor

$-r$ = Negative resisitance

(a)   (b)

Figure 1-26 *Tunnel Diode: (a) Symbol (b) Equivalent Circuit*

The use of a tunnel diode in a sinusoidal oscillator circuit is shown in Fig. 1-27. The tank circuit comprises the capacitor C_1 and inductor L_1, respectively. The resistance R_B is used to operate the tunnel diode between the peak and valley point.

The advantages of the tunnel diode are low noise, ease of operation,

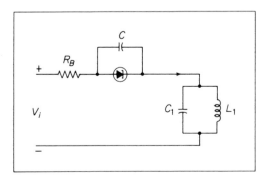

Figure 1-27 *Tunnel Diode Oscillator Circuit*

high speed, low power and temperature variation and nuclear radiation insensitivity. Some inherent disadvantages are: (1) The voltage range over which the tunnel diode can be operated properly is limited to one volt or less; (2) The diode is a two-terminal device and provides no isolation between input and output.

problems

1. A Ge diode has an I_0 equal to 50 μA at a temperature of 30°C. If the applied voltage is .15 volt, determine the operating current.

2. A Ge diode has an I_0 equal to 30 μA at a temperature of 25°C. If the applied voltage is 0.2 volt, determine the operating current.

3. A Ge diode has an I_0 equal to 40 μA at a temperature of 30°C. If the operating current is 25 mA, determine the applied voltage.

4. A Ge diode has an I_0 equal to 30 μA at a temperature of 20°C. If the operating current is 10 mA, determine the applied voltage.

5. A Ge diode has an I_0 equal to 20 μA. If the applied voltage is 0.2 volt and the operating current is 20 mA, determine the operating temperature in °C.

6. A Ge diode has an I_0 equal to 40 μA. If the applied voltage is 0.15 volt and the operating current is 10 mA, determine the operating temperature in °C.

7. A Ge diode has an operating current of 10 mA when the voltage applied is 0.2 volt at a temperature of 20°C. Determine the leakage current.

8. A Ge diode has a voltage applied of 0.18 volt at 25°C and the resulting operating current is 20 mA. Determine the leakage current.

9. A Ge diode has a leakage current of 10 μA at 20°C. The applied voltage is .1 volt at 20°C. Find:
   (a) the operating temperature for the leakage current to be equal to 35 μA;
   (b) the operating current at 20°C;
   (c) the dynamic resistance of the diode at 20°C.

10. A Ge diode has a leakage current of 15 μA at 10°C. As the temperature increases, the leakage current rises to 45 μA. Find:
    (a) the new operating temperature;
    (b) the operating current at the new temperature if the applied voltage is 0.2 volt.
    (c) the dynamic resistance at the new temperature.

11. The dynamic resistance of a Ge diode is 0.75 Ω at 30°C. The leakage current is 30 μA at 60°C. Find the applied voltage at the temperature of 30°C.

12. The dynamic resistance of a Ge diode is 1.2 Ω at a temperature of 10°C. The leakage current is 40 μA at a temperature of 30°C. Find the applied voltage at 10°C.

13. The dynamic resistance of a Ge diode is 0.5 Ω. The leakage current is 10 μA. The applied positive voltage to the diode is .15 volt. Find the operating temperature.

14. The dynamic resistance of a Ge diode is 1 Ω. The leakage current is 25 μA. The applied positive voltage to the diode is 0.2 volt. Find the operating temperature.

two

INTRODUCTION TO TRANSISTORS

PNP TRANSISTOR

The analysis of the PN junction diode characteristic indicates that forward biasing produces a low resistance element, whereas reverse bias produces a high resistance circuit element.

Assume that a crystal containing two PN junctions is prepared, one biased in the forward direction and the other junction biased in the reverse direction. A signal could be introduced into the low resistance circuit and removed from the high resistance circuit. If the same current flows through both junctions, the power gain of the system could be extremely high. Consequently, such a device can transfer a signal from the low resistance element into a high resistance element. Contracting the words *transfer resistor* forms the word *transistor*.

Arrange a single crystal with two PN junctions in the form of a sandwich as shown in Fig. 2-1. The leads are identified as the emitter, base and collector.

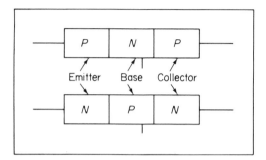

Figure 2-1 *PNP or NPN Transistor*

It is evident that either combination results in two depletion regions that occur at the junctions even though there is no application of external voltages. It is possible for either holes or electrons to flow across the barrier regions. For proper transistor action, the first, or input, junction should be biased in the forward direction and the second, or output, junction should be biased in the reverse direction.

Connect a PNP transistor with the external voltages as shown in Fig. 2-2. With the application of the voltages, the holes in the emitter region are repelled by the battery potential and begin to diffuse toward the emitter to base junction.

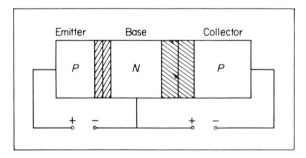

Figure 2-2 *PNP Transistor*

The forward bias potential reduces the depletion region and the majority of the holes pass through the relatively small base area (N region) into the collector region. A small number of holes are neutralized by combining with electrons in the base region. As each of the holes enters the collector region, it is neutralized by an electron emitted by the negative terminal of the battery.

For each hole neutralized by combining with an electron in the base or collector region, an electron from one of the covalent bonds near the emitter terminal lead enters the positive terminal of the battery. This results in the formation of a new hole that diffuses toward the junction to maintain a continuous flow of current from emitter to collector. In practical transistors, approximately 92 to 99 percent of the holes from the emitter reach the collector region.

The flow lines of holes and electrons within a PNP transistor are shown in Fig. 2-3.

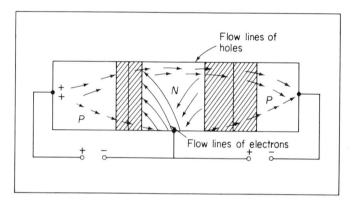

Figure 2-3 *Flow Lines in a PNP Transistor*

NPN TRANSISTOR

The theory of operation for the NPN transistor is similar to that of the PNP transistor. Consider the NPN transistor connected to external sources as shown in Fig. 2-4.

The emitter-to-base potential forward biases the transistor narrowing the depletion region and permitting electrons to diffuse across the junction barrier. Since the base region is relatively thin, the electrons will not combine with the holes in this region but will continue into the collector region. Although the collector to base is reverse biased, the electrons will diffuse across the collector junction. Once in the collector region, the electrons are attracted to the positive collector electrode. The flow lines of electrons and holes are illustrated in Fig. 2-4.

TRANSISTOR SYMBOL

The symbols used for transistors are illustrated in Fig. 2-5.

A horizontal line represents the base. The two angular lines represent

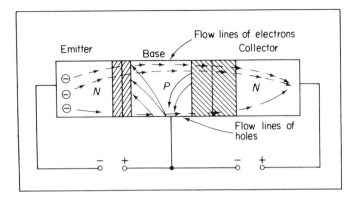

Figure 2-4   *Flow Lines in an NPN Transistor*

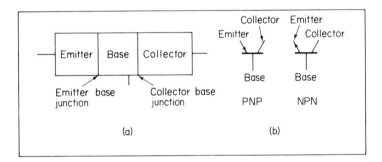

Figure 2-5   *Transistor Nomenclature and Symbols*

the emitter and collector respectively. The angular line with the arrowhead on it identifies the emitter; the other angular line represents the collector. When the arrow points toward the base, the symbol represents a PNP transistor. If the arrow points away from the base, the symbol represents an NPN transistor. The arrowhead on the transistor symbol indicates the direction of conventional current flow.

EMITTER, BASE AND COLLECTOR CURRENTS

A properly biased NPN transistor is shown in Fig. 2-6. The purpose of resistor R is to limit the emitter current I_E. The terminology used will be:

V_{EE}—emitter supply battery
V_{CC}—collector supply battery
I_E—emitter current
I_B—base current
I_C—collector current
I_{CO}—reverse bias current (leakage current)

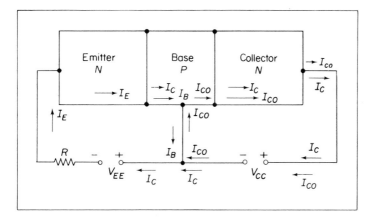

Figure 2-6   *Transistor Currents Using Electron Flow*

The application of the emitter supply voltage V_{EE} causes I_E to flow. Inside the crystal, the entire current flowing in at the emitter is carried by electrons created at the emitter terminal. When the electrons diffuse into the base region, the total electron flow splits and flows in two directions. Some 92 to 99 percent of the electrons emitted into the base reach the collector and the positive terminal of the collector supply voltage, V_{CC}.

The remainder of the injected electrons meet and combine with the holes created in the base terminal. Thus, by the mechanism of the electron pairs, part of the total current flows out of the base and back to the emitter supply voltage. This small amount of current that flows through the base terminal is the base current, designated as I_B. Hence, the emitter current is equal to the sum of the collector and base currents, or:

$$I_E = I_C + I_B$$

This equation expresses the fundamental relationship between currents in a transistor circuit.

ALPHA

Note that a large percentage of the total emitter current survives the trip through the base and appears at the collector terminal. The remainder of the emitter current is diverted through the base terminal. Consequently, the collector current can be represented as the sum of the leakage current, designated I_{co}, and that percentage of the emitter current that reaches the collector, or:

$$I_C = I_{co} + \text{percentage of } I_E$$

The term that defines that percentage is given the name *alpha* and designated by the symbol α. The collector current is expressed by:

$$I_C = I_{co} + \alpha I_E$$

The base current can also be expressed in terms of the emitter current (I_E) and alpha (α). Thus,

$$I_B = I_E - I_C$$

or

$$I_B = I_E(1 - \alpha) - I_{co}$$

Actually, currents cannot flow in opposite directions as shown in the base lead, but the leakage current reduces the flow of base current.

CHARACTERISTIC CURVES

A complete description of the electrical circuit behavior of the transistor can be readily established by construction of the static characteristics of the device. Note that in Fig. 2-7, the common return for the source voltages is the base; therefore, it is called the common or grounded base circuit. There are inherently four variables to contend with. These are:

    1. The base to emitter voltage (V_{be})

    2. The input current or emitter current (I_E)

    3. The collector to base voltage (V_{cb})

    4. The output collector current (I_C)

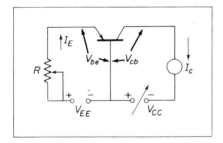

Figure 2-7 *Common Base Circuit*

The output or collector characteristics are obtained in the following manner. The input current, I_E, is varied by adjusting the potentiometer, R. The collector to base voltage is varied by adjusting the collector supply voltage, V_{CC}. A complete family of output characteristic curves is obtained and is shown in Fig. 2-8.

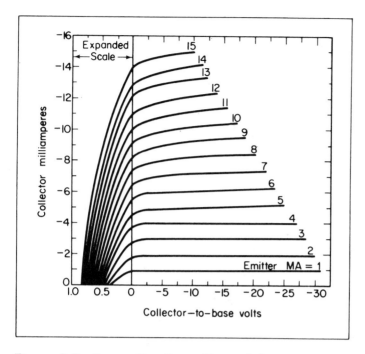

Figure 2-8 *Common Base Output Characteristic*

Note that all curves appear as straight lines. An increase in collector current occurs only when the emitter current is increased. At any point on the graph the collector current is slightly less than the emitter current causing it.

EMITTER FAMILY OF CURVES

In the common emitter circuit, the emitter terminal is common to both the input and output circuit as shown in Fig. 2-9. Bias is applied accord-

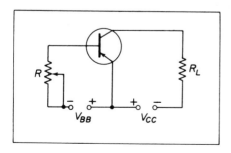

Figure 2-9 *Common Emitter Circuit*

ing to the basic rules previously specified. The emitter to base is forward biased. Usually, V_{CC} is much larger than V_{BB}, consequently the collector to base is reversed biased. The resistor, R, is a current-limiting resistor and provides the proper base bias current. The complete output characteristic is shown in Fig. 2-10.

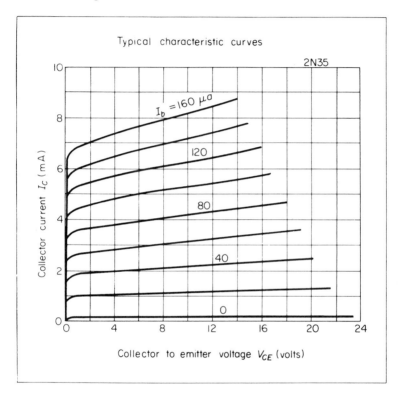

Figure 2-10 *Typical Characteristic Curves*

The characteristic curve labeled $I_b = 0$ represents the leakage current flowing from the emitter to collector with the base terminal open. The mathematical expressions existing in the circuit are:

$$I_C = \alpha I_E + I_{co}$$

and

$$I_E = I_C + I_B$$

Substituting for I_E yields

$$I_C = \alpha(I_C + I_B) + I_{co}$$

then

$$I_C = \frac{\alpha}{1 - \alpha} I_B + \frac{\alpha}{1 - \alpha} I_{co}$$

The ratio $\dfrac{\alpha}{1 - \alpha}$ is called *beta* and is designated by the Greek letter β. Thus,

$$\beta = \frac{\alpha}{1 - \alpha}$$

and

$$\beta + 1 = \frac{1}{1 - \alpha}$$

Consequently, the equation for I_C is:

$$I_C = \beta I_B + (\beta + 1) I_{co}$$

The fact that leakage current is multiplied by a factor of beta plus one denotes that the common emitter circuit is sensitive to temperature variations.

GRAPHICAL ANALYSIS

Transistors require certain supply and bias voltages to set the correct operating conditions. Bias is provided in transistor circuits so that the point of operation lies midway between the limits of *cutoff* and *saturation*.

These limits are shown in Fig. 2-11. At the cutoff point I_B is equal to zero. To bring the leakage current down, the input circuit should actually be reverse biased. Saturation occurs below the knee of the curves. In this region, the steep, almost vertical characteristic shows that the transistor is acting as a very small resistance. Between cutoff and saturation is the active amplifying region of operation. In this region, the emitter to base is forward biased and the collector is reverse biased.

The graphical analysis is used to construct the dc load line. A load line graphs all possible coordinates of the collector to emitter voltage (V_{CE}) and collector current I_C for a given collector supply voltage and load resistance.

The required system equations are for Fig. 2-9:

$$V_{ce} = V_{cc} - i_c R_L$$

The load line is constructed by using the slope intercept method. The required two points are:

Let $V_{ce} = 0$ and $i_c = \dfrac{V_{cc}}{R_L}$

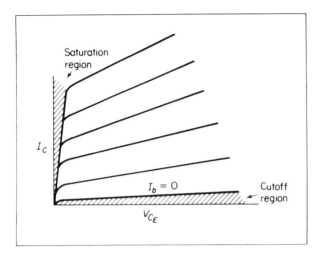

Figure 2-11 *Characteristics Showing Cutoff and Saturation Regions*

then $\qquad i_c = 0 \qquad V_{ce} = V_{cc}$

These two points are connected and the load line has been constructed. An illustrative problem will demonstrate the procedure.

sample problem

A common emitter circuit is shown below. Construct the load line.

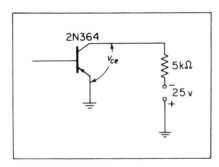

Solution:

Step 1: Determine the basic operating equation

$$V_{ce} = V_{cc} - i_c R_L$$

The first point is: $V_{ce} = V_{cc} = 25$ volts

when $i_c = 0$

Step 2:   The second point is: $i_c = \dfrac{V_{cc}}{R_L} = 5$ mA

when $V_{ce} = 0$

The load line construction is shown in Fig. 2-12.

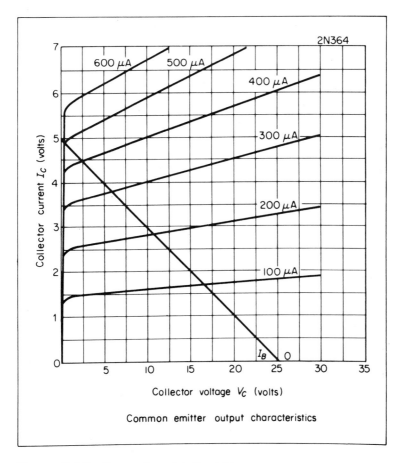

Figure 2-12 *Construction of dc Load Line*

Having constructed the load line for a given emitter circuit, the zero signal operating point must be selected. This point determines the base bias current required for proper circuit operation.

DYNAMIC TRANSFER CHARACTERISTIC

The load line constructed on the static characteristic illustrates the behavior of the circuit. Another type of characteristic curve that may simplify the explanation of circuit performance is known as the dynamic transfer characteristic. It is generally much simpler to illustrate the influence of the base signal input current on the behavior of the output collector current by means of the dynamic transfer curve.

The procedure for the construction of the dynamic transfer characteristic is illustrated by a step-by-step method. Consider the load line constructed on the characteristic shown in Fig. 2-13a. The transfer static charac-

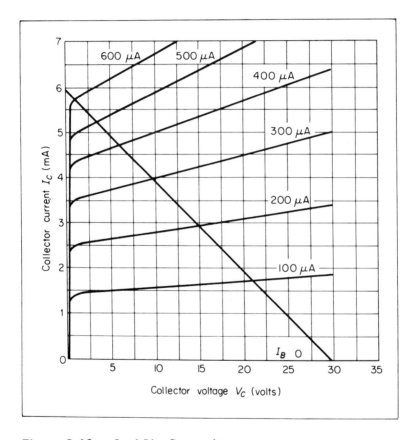

Figure 2-13a   *Load Line Construction*

teristic is shown in Fig. 2-13b for a constant V_{ce} equal to 10 volts. The two families of curves have three quantities in common:

1. A common collector current axis
2. Like values of base current
3. Like values of collector voltage

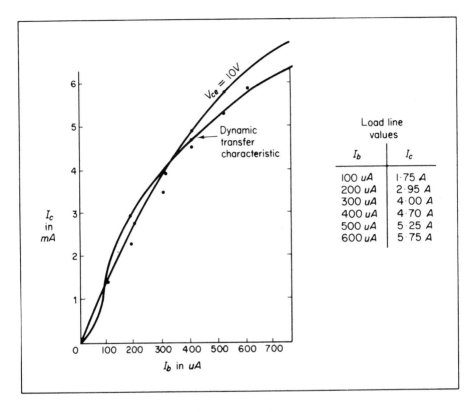

Figure 2-13b *Dynamic Transfer Characteristic*

To project the effect of the load line on the transfer characteristic, it is necessary to establish a collector current for each value of the base current, using the following procedure:

Step 1: Determine the value of I_c on the original load line using $I_b = 100 \ \mu A$. The reading from the graph should be 1.75 A.

Step 2: Place this value on the transfer static curve.

Step 3: Determine the value of I_c at $I_b = 200 \ \mu A$. From the graph, the reading should be 2.95 A.

Step 4: Again, this value is placed on the static transfer characteristics.

Step 5: Continue this procedure until a minimum of five or six points have been determined. Place these values on the static transfer characteristic curve.

Step 6: Connect these points sequentially to establish the dynamic transfer characteristic.

The curve shown on the static transfer characteristic was obtained holding V_{ce} at 10 volts under static conditions. The collector voltage of the dynamic transfer characteristic curve is not shown as a constant since the collector voltage varies due to the load resistor producing this variation.

The purpose of the dynamic transfer characteristic is to provide a simplified method to establish a proper point of operation for the transistor. Thus, for linear operation, the operating point is established within the linear portion of the dynamic transfer characteristic.

The location of the operating point and the amplitude of the input signal will determine whether the transistor is functioning properly. Proper and improper points of operation are illustrated in Fig. 2-14a and b, respectively.

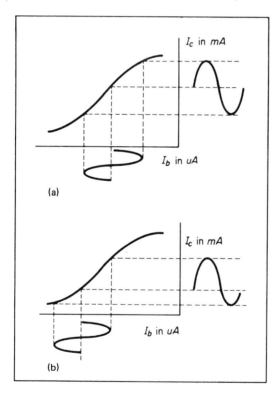

Figure 2-14 (*a*) *Proper Point of Operation for Signal Input* (*b*) *Improper Point of Operation Producing Distorted Signal Output.*

BIASING TECHNIQUES

Various methods have been established for biasing the base circuit in a common or grounded emitter circuit. In Fig. 2-15, two separate and dis-

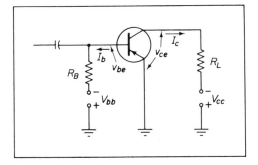

Figure 2-15   *Two-Battery Fixed Bias System*

tinct bias or battery supply voltages are used. This system is called the two-battery fixed bias system. The performance of the bias circuit is established by the equation:

$$v_{be} = V_{bb} - I_b R_B$$

Generally, v_{be} is an exceedingly small voltage compared to the base supply voltage, so the equation simplifies to:

$$I_b = \frac{V_{bb}}{R_B}$$

It should be noted that both the base and collector are negative with respect to the emitter. Therefore, a single battery can be used to supply the fixed bias current for the base circuit. This method is illustrated in Fig. 2-15a.

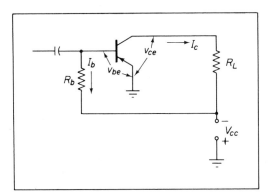

Figure 2-15a   *Single-Battery Fixed Bias Circuit*

Using the assumption that the value of v_{be} is negligible, the base current I_b is given by:

$$I_b = \frac{V_{CC}}{R_B}$$

sample problem

A common emitter circuit is shown in the figure below. Construct the load line and determine the point of operation.

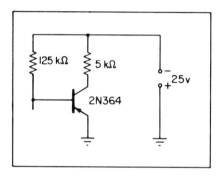

Solution:

Step 1:   Determine the basic operating equation.

$$V_{ce} = V_{cc} - i_c R_L$$

The first point is: $V_{ce} = V_{cc} = 25$ volts when $i_c = 0$

Step 2:   The second point is:

$$i_c = \frac{V_{CC}}{R_L} = 5 \text{ mA when } V_{ce} = 0$$

Step 3:   The base bias current is defined by:

$$i_b = \frac{V_{CC}}{R_B} = \frac{25}{125 \times 10^3} = 200 \ \mu\text{A when } v_{be} = o$$

Step 4:   The point of intersection of $i_b = 200 \ \mu$A and the *dc* load line defines the point of operation.

The load line and point of operation are shown in Fig. 2-16.

Fixed bias is not the most satisfactory method for biasing the base circuit. Unit to unit variations among transistors and their extreme sensitivity to temperature variations make it difficult to establish a critical value of base bias current.

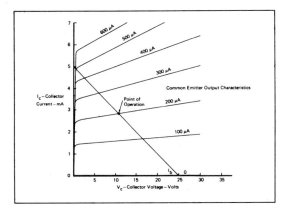

Figure 2-16

One method of correcting this situation is to connect the base resistor to the collector as shown in Fig. 2-17. This type of self bias arrangement has the disadvantage of feeding back or returning the output signal to the input circuit. The value of the base bias resistor is found by:

$$v_{be} = V_{ce} - i_b R_B$$

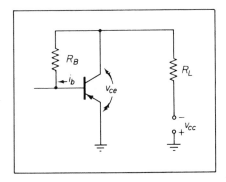

Figure 2-17 *Self Bias Circuit*

Using the assumption that the value of v_{be} is negligible, the base current i_b is:

$$i_b = \frac{V_{ce}}{R_B}$$

Both fixed and self bias can be used to provide better circuit stability. This method is illustrated in Fig. 2-18.

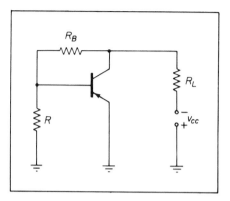

Figure 2-18   *Combination Bias Circuit*

In this circuit a voltage divider circuit composed of R_B and R biases the base circuit negative with respect to the emitter. Bleeder current flowing through the voltage divider determines the base bias current. Note that any variation in the collector voltage produces an instantaneous variation in the base bias current.

A more useful circuit introduces an emitter resistor and connects R_B to the negative battery terminal, as shown in Fig. 2-19.

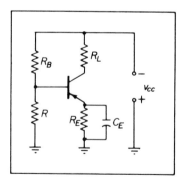

Figure 2-19   *Emitter Resistor Base Bias Circuit*

The emitter resistor, R_E, provides additional stability and varies in ohmic resistance from 1/5 to 1/10 the value of R. To prevent emitter degeneration, capacitor C_E is provided across R_E. The value of this capacitor varies from 50 μF to 1000 μF depending upon the lowest frequency of interest.

AC LOAD LINES

In practical amplifiers, the load on the transistor is not a simple resistor. Consider the circuit shown in Fig. 2-20. The first step in the analysis of the system is to construct the dc load line. The slope intercept method is used. Thus,

let $i_c = 0$ and $V_{ce} = V_{cc}$

$$V_{ce} = 0 \qquad i_c = \frac{V_{cc}}{R_L}$$

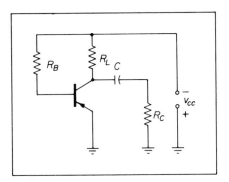

Figure 2-20   *AC Load Line Circuit*

The next step in the procedure of constructing an ac load line is to determine the operating point. The required equation is:

$$i_b = \frac{V_{cc}}{R_B}$$

This value is placed on the dc load line.

The operating point is a common point for both the dc and ac load lines. The ac load resistance is measured between collector and ground. In this case, the coupling capacitor, C, is assumed to be a short circuit. Thus,

$$R_{L_{ac}} = \frac{R_L R_C}{R_L + R_C}$$

The construction of the ac load line proceeds as follows. $R_{L_{ac}}$ is held constant. Assume a voltage applied across $R_{L_{ac}}$ and calculate the current. Using the operating point as one point on the ac load line, a second point is established by the above calculation and placed on the graph in the following manner. From the operating point, lay off the assumed value of voltage parallel to the V_{ce} axis. At the end of the voltage vector, construct a vertical line equal to the calculated current. This defines the second point. Connect this point with the operating point and extend line in both directions. The ac load line has been constructed.

An illustrative problem will demonstrate the theory.

sample problem

A common emitter circuit is shown below. Determine the operating point and construct the ac load line.

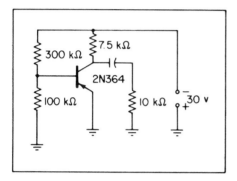

Solution:

Step 1:   Construct the dc load line. Using the slope intercept method:

Let $V_{ce} = 0$ $i_c = 0$

then $i_c = \dfrac{30}{7.5} 10^{-3} \doteq 4 \text{ mA}$ $V_{ce} = 30 \text{ V}$

Step 2:   Determine the base bias current. The base bias circuit is Thevenized as shown below.

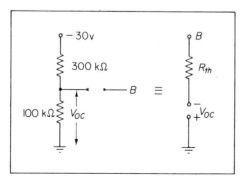

$$V_{OC} = 30\frac{100}{400} = 7.5 \text{ V}$$

$$R_{th} = \frac{300 \times 100}{400} \text{ K} = 75 \text{ K}$$

Step 3: Assuming $V_{be} = 0$, the base bias current is given by:

$$i_b = \frac{V_{CC}}{R_{th}} = 100 \text{ } \mu\text{A}.$$

This value is placed at point of intersection of $i_b = 100 \text{ } \mu\text{A}$ line and the dc load line and defines the operating point.

Step 4: Calculate $R_{L_{ac}}$

$$R_{L_{ac}} = \frac{7.5 \times 10}{7.5 + 10} \text{ K}$$

$$R_{L_{ac}} = 4.29 \text{ K}$$

Step 5: Assume an applied voltage across $R_{L_{ac}}$ and calculate the current. Thus,

$$E_{\text{assumed}} = 4.3 \text{ V}$$

$$I_{\text{cal}} = 1 \text{ mA}$$

The construction of the ac load line is shown in Fig. 2-21.

A resistor is inserted in series with the emitter to form a self biasing circuit. Consider the circuit shown in Fig. 2-22.

The following procedure is used to construct the ac load line. The first step is to construct the dc load line using the relationship

$$V_{cc} = I_c R_L + V_{ce} + I_E R_E$$

and since

$$I_E = I_C + I_B$$

then $\quad V_{CC} = I_C(R_L + R_E) + V_{ce} + I_B R_E$

Assuming $I_B R_E$ is negligible, then the value for I_C is:

$$I_C = \frac{V_{CC} - V_{ce}}{R_L + R_E}$$

Note: C_C and C_E are considered open circuits for dc. X_{CC} and X_{CE} are considered short circuits for ac.

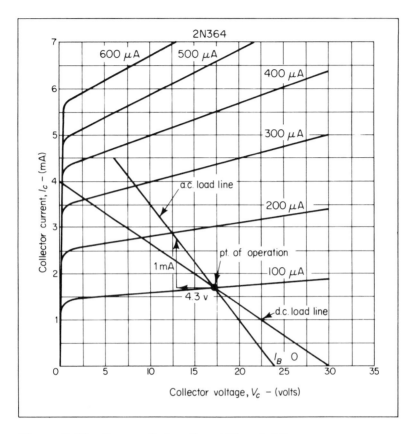

Figure 2-21 *Common Emitter Output Characteristics*

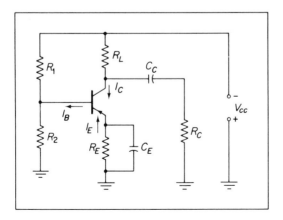

Figure 2-22 *Emitter Self Biasing Circuit*

Thus, using the slope intercept method:

Let $V_{ce} = 0$ and $I_c = \dfrac{V_{CC}}{R_L + R_E}$

$I_c = 0$ and $V_{ce} = V_{CC}$

Place these points on the collector characteristics and the dc load line has been constructed. The next step in the procedure is to determine the point of operation. Thevenize the base circuit. Thus, refer to Fig. 2-23.

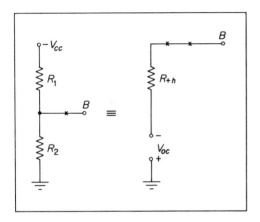

Figure 2-23

$$V_{OC} = V_{CC} \frac{R_2}{R_1 + R_2}$$

$$R_{th} = \frac{R_1 R_2}{R_1 + R_2}$$

The equivalent circuit is used to determine the base circuit equation. Thus,

$$V_{OC} = I_B(R_{th} + R_E) + I_C R_E$$

Substituting for I_C yields:

$$I_B = \frac{V_{OC} - \left(\dfrac{V_{CC} - V_{ce}}{R_L + R_E}\right) R_E}{R_{th} + R_E}$$

The unknowns in the preceding equation are I_B and V_{ce}. For each value of V_{ce}, a corresponding value of I_B exists defining a point on the bias curve. These points are then connected. The intersection of the bias curve and the load line specifies the point of operation. Once the point of operation has been defined, the ac load line is constructed as previously. Thus,

$$R_{L_{ac}} = \frac{R_L R_c}{R_L + R_c}$$

Note that emitter R_E is bypassed by C_E. Since X_{CE} is approximately zero at the lowest frequency of interest, it shorts out R_E.

An illustrative problem will demonstrate the theory.

sample problem

Given the circuit shown below, determine the dc load line point of operation, and the ac load line.

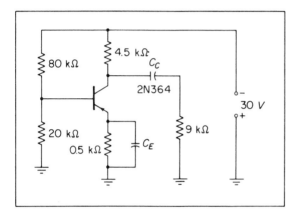

Solution:

Step 1: Determine the dc load line

$$V_{ce} = V_{CC} - I_c(R_L + R_E)$$

Let $I_c = 0$ and $V_{ce} = V_{CC} = 30$ V

$$V_{ce} = 0 \text{ and } I_C = \frac{30}{5 \text{ K}} = 6 \text{ mA}$$

Place these values on the 2N 364 characteristics and construct the dc load line.

Step 2: Thevenize the base circuit. Determine V_{OC} and R_{th}.

$$V_{OC} = V_{ce}\frac{R_2}{R_1 + R_2}$$

$$V_{OC} = 6 \text{ volts}$$

$$R_{th} = \frac{R_1 R_2}{R_1 + R_2}$$

$$R_{th} = 16 \text{ K}$$

Step 3: Construct the bias curve.

$$I_b = \frac{V_{oc} - \left(\frac{V_{cc} - V_{ce}}{R_L + R_E}\right) R_E}{R_{th} + R_E}$$

V_{ce}	I_B
10	243 μA
15	272 μA

Place these points on the characteristic curves and the point of intersection with the dc load line defines the point of operation.

Step 4: Calculate $R_{L_{ac}}$

$$R_{L_{ac}} = \frac{R_L R_C}{R_L + R_C}$$

$$R_{L_{ac}} = 3 \text{ K}$$

Step 5: Assume an applied voltage and calculate the current. Thus, let $V_{ac} = 9$ volts. Then $I_{ac} = 3$ mA. The construction is shown in Fig. 2-24.

With the ac load line constructed, it is possible to determine the variation in the operating conditions with the application of an input signal. Since the transistor is a *current activated* device, assume a sinusoidal input current as shown in Fig. 2-25. Note that the base current swings 100 μA in each direction along the ac load line. Two new points are shown located at points A and B, respectively. Note the corresponding changes in both I_C and V_{ce}. The peak to peak collector current and voltage swing can be evaluated from Fig. 2-25.

The current amplification of the device is defined as the ratio of the output collector current to the input base current swing. Thus,

$$A_i = \frac{I_C(\text{peak to peak swing})}{I_B(\text{peak to peak swing})}$$

In this case, both currents are given in peak to peak values and the current amplification is equal to:

$$A_i = \frac{(4.15 - 1.75) \text{ mA}}{200 \ \mu\text{A}}$$

$$A_i = 12$$

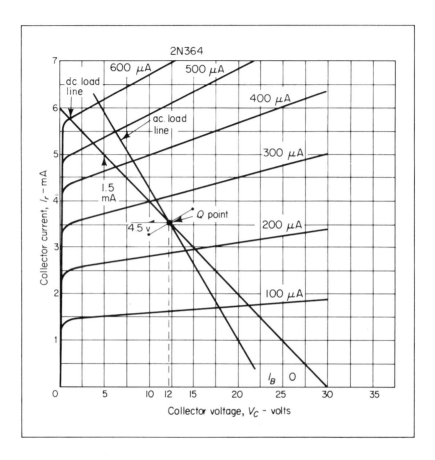

Figure 2-24 *Construction of AC Load Line*

The voltage amplification of the device is defined as the ratio of the output collector voltage to the input base voltage. Thus,

$$A_v = \frac{V_{ce} \text{ (peak to peak swing)}}{V_{be} \text{ (peak to peak swing)}}$$

Using the collector characteristics of Fig. 2-25, the collector voltage swing is given as:

$$V_{ce} = 18.75 - 11.5 = 7.25 \text{ V}$$

The base to emitter signal input voltage may be assumed about .25 volt peak to peak swing. The voltage amplification, then, is:

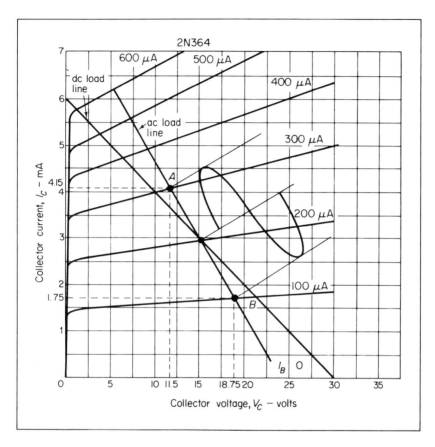

Figure 2-25   *Sinusoidal Input on AC Load Line*

$$A_v = \frac{7.25}{.25}$$

$$A_v = 29$$

The power gain is given by the product of the current gain and the voltage gain. Thus,

$$P_g = A_v A_i$$

BIAS STABILIZATION

The operating point of a transistor circuit establishes the zero signal condition of the transistor. Bias stabilization deals with the *control* of the operating point bias to prevent excessive shift from the preset value. This shift can occur because of transistor unit to unit variation or temperature variations.

It is evident that the semiconductor leakage current and junction potential are extremely sensitive to temperature changes. The effects of a change in leakage current on the Q point are shown in Fig. 2-26.

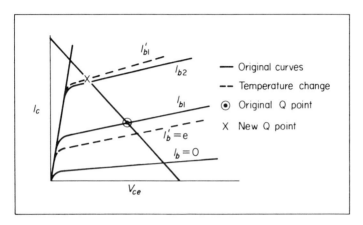

Figure 2-26 *Characteristic Curves with Temperature Variation*

An increase in temperature raises the leakage current and causes all the curves to be raised upward as shown in the dotted lines of Fig. 2-26. As a result, the Q point moves upward and the signal handling ability of the amplifier is impaired.

Another temperature effect that occurs in transistors is known as *thermal runaway*. An external temperature increase raises the leakage current. This increased leakage current increases the collector current. A higher collector current increases the power dissipated in the collector junction and raises the temperature. This sequence is repetitive and increases the collector current each cycle. This thermal runaway cycle can damage a transistor rapidly.

STABILITY FACTOR

In order to compare various biasing methods, a stability factor symbolized by the letter S is used. The stability factor is defined as the change in collector current to the change in collector leakage current. Thus,

$$S = \frac{\Delta I_c}{\Delta I_{co}}$$

If the stability factor is given as 30, and the change in leakage current is 100 μA, the change that would occur in the collector circuit current is given by:

$$\Delta I_C = 30 \times 100 \ \mu A = 3 \ mA$$

Consequently, it is evident that the ideal stability factor is unity. A higher stability factor represents a relatively less stable bias point operation.

At this point, it would be advisable to examine a number of different biasing circuits and evaluate the stability factor for each network.

The simplest possible biasing circuit is shown in Fig. 2-27.

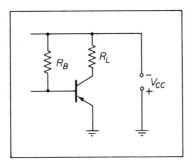

Figure 2-27 *Fixed Bias Circuit*

Note that current I_B is given by:

$$I_B = \frac{V_{CC} - v_{be}}{R_B}$$

Assuming that the voltage v_{be} is negligible, the relationship for I_B becomes

$$I_B = \frac{V_{CC}}{R_B}$$

The collector current is given by:

$$I_C = \beta I_B + (\beta + 1) I_{co}$$

Consequently, substituting for I_B, the stability factor for this circuit is given by:

$$S = \beta + 1$$

This is an extremely poor bias stable circuit.

An improved stabilization method is shown in Fig. 2-28. The applied

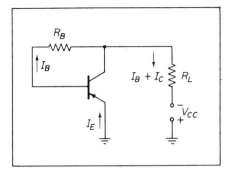

Figure 2-28   *Improved Bias Circuit*

voltage to the base bias resistor is the collector to emitter voltage. The following equations are applicable.

analysis

$$V_{ce} = V_{CC} - I_C R_L - I_B R_L$$

and assuming that v_{be} is negligible, the value for I_B is:

$$I_B = \frac{V_{ce}}{R_B} = \frac{V_{CC} - I_C R_L}{R_B + R_L}$$

Substituting and simplifying yields the value for I_C:

$$I_C = \beta \left(\frac{V_{CC} - I_C R_L}{R_B + R_L} \right) + (\beta + 1) I_{co}$$

and

$$I_C = \frac{\dfrac{\beta V_{CC}}{R_B + R_L} + (\beta + 1) I_{co}}{\dfrac{\beta R_L}{R_B + R_L} + 1}$$

The stability factor is given by:

$$S = \frac{(\beta + 1)(R_E + R_B)}{(\beta + 1) R_E + R_B}$$

A further improvement in stabilization of the base bias circuit is shown in Fig. 2-29. Note that resistor R_E is placed in series with the emitter lead. The direction of current flow through the emitter resistor indicates that the voltage at the emitter terminal is negative with respect to ground.

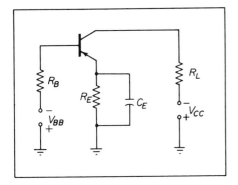

Figure 2-29   *Emitter Bias Circuit*

The base to ground voltage must be greater than the emitter to ground voltage for the junction to be forward biased.

analysis

$$I_C = (\beta + 1) I_{co} + \beta I_B$$

$$V_{BB} = I_B R_B + \overset{\text{negligible}}{\cancel{v_{be}}} + I_E R_E$$

Then

$$I_B = \frac{V_{BB} - I_C R_E}{R_E + R_B}$$

Substituting yields

$$I_C = \beta \left(\frac{V_{BB} - I_C R_E}{R_E + R_B} \right) + (\beta + 1) I_{co}$$

$$I_C = \frac{\dfrac{\beta V_{BB}}{R_B + R_E} + (\beta + 1) I_{co}}{1 + \dfrac{\beta R_E}{R_B + R_E}}$$

The stability factor is given by:

$$S = \frac{(\beta + 1)(R_B + R_E)}{R_B + R_E (1 + \beta)}$$

A practical circuit arrangement used for the base bias improved stability circuit is shown in Fig. 2-30. The analysis of this circuit is identical to the circuit analysis developed for the circuit given in Fig. 2-29. The values of V_{BB} and R_B are found by Thevenizing the base circuit as shown in Fig. 2-31.

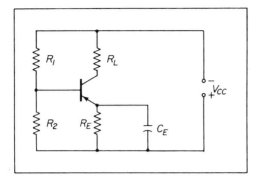

Figure 2-30 *Feedback Bias Circuit*

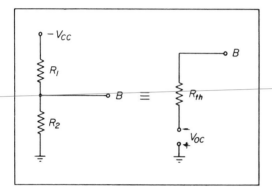

Figure 2-31 *Thevenizing the Base Circuit*

Note that V_{BB} is equal to V_{OC} and is defined by:

$$V_{OC} = V_{CC} \frac{R_2}{R_1 + R_2}$$

and R_B is equal to R_{th} and is defined by:

$$R_{th} = \frac{R_1 R_2}{R_1 + R_2}$$

NONLINEAR COMPENSATION METHODS

The bias stabilization methods previously discussed acted to maintain the collector current point of operation relatively constant. A modern technique utilizes components that vary with temperature in the same manner as transistors. Consider the circuit shown in Fig. 2-32. A semicon-

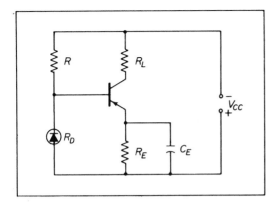

Figure 2-32   *Semiconductor Diode Used for Bias*

ductor diode is used to provide the base bias voltage and current. This diode is made of the same materials as the transistor and has the identical temperature characteristics.

As the temperature rises, the voltage drop across the diode decreases, reducing the transistor forward bias at the required rate. Since the forward bias impedance of the diode is extremely low, base current variations produce negligible variations across it.

Thermistors are resistors made from semiconductor materials. The forward bias for a transistor utilizes a thermistor as shown in Fig. 2-33. As the temperature rises, the resistance of the thermistor decreases, reducing the voltage drop across it. This also decreases the forward bias. Most thermistors have a higher negative temperature coefficient of resistance than germanium. The use of the thermistor in the circuit by itself would reduce the forward bias too rapidly.

Consequently, the thermistor is shunted by the fixed resistor, R. The adjustment of the resistor R to a required preset value permits the bias

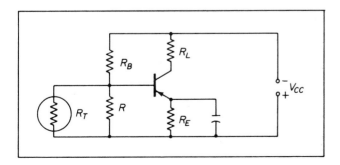

Figure 2-33   *Thermistor Bias Stabilization*

to vary accordingly. Note that the purpose of the temperature sensitive element is to compensate for the temperature variations in the semiconductor junction. Therefore, the temperature sensitive element should be located as close as possible to the transistor case.

THE JFET

The field effect transistor, abbreviated FET, is a semiconductor device that acts similar to a vacuum tube in that the charge carriers are controlled by an electric field at the control electrode. The typical PNP- or NPN-type transistor operates on two PN junctions and consequently is called a *bipolar junction device*. The field effect transistor is considered *unipolar*, utilizing only the free majority carriers in the conducting region.

The structure of the N channel junction field effect transistor (JFET) is shown in Fig. 2-34.

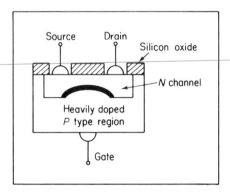

Figure 2-34   *Construction of JFET*

The two end terminals of the N channel are metal to semiconductor ohmic contacts and are referred to as the source (S) and drain (D), respectively. Between the source and drain terminals is the resistance of the N region. A gate (G) is formed by placing a ring of heavily doped P-type material around the center of the channel to produce a PN junction. When reverse bias is applied between the gate and source, the depletion zone widens. An examination of the conduction path between the source and drain indicates that the ohmic resistance increases as the depletion region widens.

The source is the electrode from which the majority carriers are emitted to flow toward the drain. The drain is the electrode that collects these majority carriers at the ohmic contact. The gate is the electrode that controls the flow of the majority carriers between the source and drain. The channel is the narrow region through which the majority carriers flow from the source to the drain.

theory of operation   The cross sectional view of a JFET is illustated in Fig. 2-35a. It is evident that a PN junction exists between the gate and the channel with its own depletion region. Since the operation of the JFET does not depend in any way on this depletion region existence, the PN junction will be ignored.

The gate is normally reverse biased with respect to the source as shown in Fig. 2-35b. The leakage flow is zero. Increasing the negative voltage on the gate with respect to the source will increase the size of the depletion region and reduce the drain current flow. As the gate bias voltage is further increased, the drain current becomes smaller and smaller until a condition of cutoff is reached. This condition of cutoff is shown in Fig. 2-35c.

The symbol for the JFET is shown in Fig. 2-36. The direction of the arrow denotes the direction in which the gate current flows if the gate is forward biased.

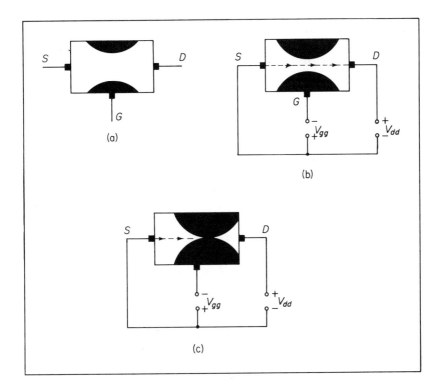

Figure 2-35   *The Electric Field in the JFET: (a) Cross Section (b) Normal Operation (c) Cutoff Operation*

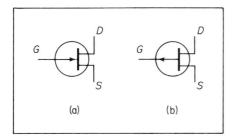

(a)   (b)

Figure 2-36   *Symbol for the JFET* (a) *N Channel* (b) *P Channel*

static characteristics   A circuit used to obtain static characteristics for the
FET is shown in Fig. 2-37. The typical static characteristics are shown in
Fig. 2-38. Note that the *VI* characteristic is a plot of drain current versus
drain voltage with the gate to source voltage (V_{gs}) as a parameter. Consider
the case when the gate voltage is at zero voltage.

In response to the applied V_{ds}, the N channel bar acts as a resistor to
the current I_d and increases linearly. As the current increases, the ohmic
voltage drop existing between the source and the channel region reverse
biases the junction and the conducting portion of the channel narrows. This
narrowing is nonuniform, wide at the source and narrowing tremendously
as the drain electrode is approached.

The basic rules for connecting source voltages to the JFET are:

1. The gate to source electrodes are reverse biased

2. The drain to source electrodes are also biased for majority carrier
   flow from S to D

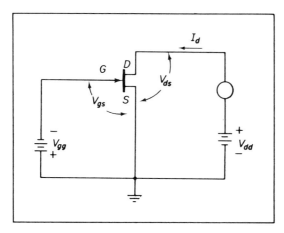

Figure 2-37   *Common Source JFET Circuit*

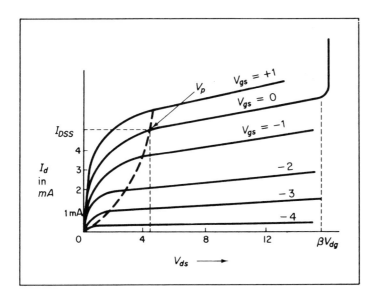

Figure 2-38   *The Static Characteristics of an N Channel FET Illustrating*
(a) Pinch-off (Vp) (b) Breakdown Voltage

The source electrode is generally considered the reference electrode
or ground point.

Examination of the static characteristics indicates that for low values
of V_{ds}, the drain current increases linearly. For values of V_{ds} greater than
the pinch-off voltage (V_p), the drain current is comparatively independent
of V_{ds}. For any given FET, the value of pinch-off voltage may be determined from the knee of the $V_{gs} = 0$ curve. A rule of thumb estimate for
evaluating V_p is to use the value of V_{gs} that reduces the drain current (I_d)
to approximately zero, as illustrated in Fig. 2-38.

The voltage V_{ds} can be further increased until the reverse voltage
breakdown point is reached. On the $V_{gs} = 0$ curve, the breakdown voltage
is symbolized as βV_{dg} and refers to the breakdown voltage between the drain
and gate with the source electrode open circuited.

load lines   Consider the typical JFET circuit shown in Fig. 2-39.

The basic circuit equations can be written

$$V_d = V_{DD} - I_d(R_L + R_s)$$
$$V_{gs} = -I_d R_s$$

The construction of the load line is identical to the method used
previously for the bipolar junction transistor. Refer to the *V-I* characteristic
shown in Fig. 2-40.

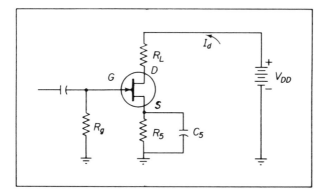

Figure 2-39   *FET Circuit*

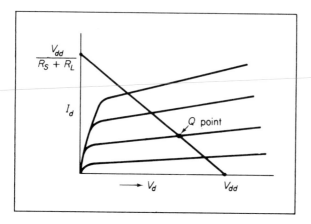

Figure 2-40   *V-I Characteristic for Load Line Construction*

Using the slope intercept method, the two points are given by:

$$\begin{cases} I_d = 0 \quad V_d = V_{DD} \\ V_d = 0 \quad I_d = \dfrac{V_{DD}}{R_L + R_s} \end{cases}$$

Place these two points on the *V-I* characteristic and construct a line connecting these two points. This line is known as the load line. A Q point or point of operation is normally assumed approximately midway between $V_{gs} = 0$ and V_{gs} equal to cutoff. The procedure for construction of the ac load line is identical to that of the bipolar junction transistor. Note that the ac load line is merely R_L since R_s is shorted out by the low ohmic resistance of X_{c_s}.

INSULATED GATE FIELD EFFECT TRANSISTOR (IGFET)

With the advent of the junction field effect transistor, it was determined that the conductivity near the surface of a semiconductor can be altered by applying a voltage to the gate metal layer that is insulated from the semiconductor. Consequently, in this type of device, the metallic gate is electrically insulated from the semiconductor surface by a thin layer of silicon dioxide.

The structure of a metal oxide semicondutor field effect transistor (abbreviated MOSFET) is shown in Fig. 2-41. Note that it consists of a thinly doped P substrate into which two heavily doped N regions are diffused. The gate structure is simply a metallic plate having no P or N type semiconductor property.

The insulation of the gate from the semiconductor results in an exceedingly high input resistance (approximately 10^{12} to 10^{15} ohms). It is evident that the gate and channel form a capacitor in which the oxide layer acts as the dielectric of a capacitor.

Because of the structural variations inherent between the depletion type MOSFET and the enhancement type MOSFET, characteristics are produced that can be utilized for separate specific circuit application.

The FET is called *unipolar* because it utilizes only one type of current carrier commonly known as majority carriers. These majority carriers are the electrons in the N channel FETs and holes in the P-type FETs. Similarly, the NPN- or PNP-type transistor is called *bipolar* because these types use both majority and minority current carriers to operate properly.

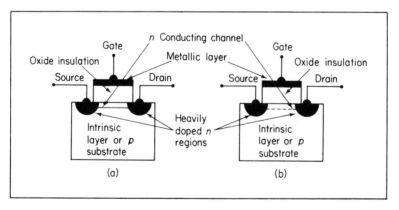

Figure 2-41   (a) *Depletion Type MOSFET* (b) *Enhancement Type MOSFET*

DEPLETION TYPE FET

Refer to the diagram having the solid or continuous line for the channel in the schematic of Fig. 2-41(a). The continuous line for the N channel denotes a *normally on* condition. The depletion region MOSFET is normally operated by making the gate voltage negative with respect to the source for an N channel MOSFET. This operating bias causes the charge carriers to be *depleted* from the conducting channel.

This mode of operation is similar to that of the JFET. Note that conduction in the channel can be completely cut off if the gate voltage is made sufficiently negative. The characteristic curves showing the normal operating region for this type of MOSFET are shown in Fig. 2-42(a).

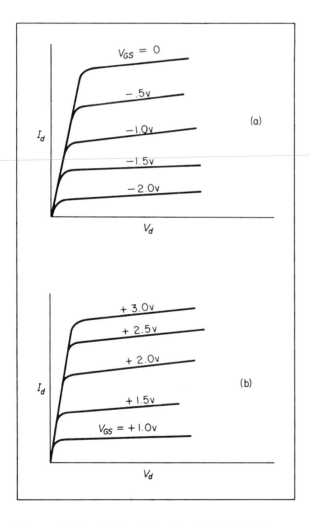

Figure 2-42   *(a) Depletion Type FET (b) Enhancement Type FET*

ENHANCEMENT TYPE FET

Refer to the diagram 2-41(b). Note that the dotted line in the N channel region specifies the enhancement type semiconductor. The enhancement region is characterized by a normally *open channel* for zero or reverse gate bias operation. (Gate voltage is made positive with respect to the source for an N channel FET or forward biased to produce the charge carriers that are drawn into the channel region.) As the forward bias is further increased, the channel conductivity between source and drain is also increased.

Typical characteristics illustrating the normal operation of the enhancement type FET are shown in Fig. 2-42(b). This type of MOSFET is well adapted for usage in digital circuitry and switching applications.

In the practical operation of any MOSFET, it is necessary to prevent stray or static voltages from being applied to the gate, or the silicon dioxide layer between the gate and the channel will be destroyed. Consequently, the grounding rings on the MOSFET must be maintained until the elements have been properly connected to the circuit and then can safely be removed without damage to the semiconductor.

The MOSFET has the advantage that there is very low leakage current because of the fact that the input resistance is very high or in the order of 10^{15} ohms. In addition, the MOSFET has the inherent characteristic for rapid switching speed of operation.

symbols The circuit symbols specified by each manufacturer for the MOSFET can vary accordingly. Some typical symbols are shown in Fig. 2-43. In general, the MOSFETs used in practical circuits have the substrate lead connected to the source internally. The circuit symbol used to represent the JFET will also represent the MOSFET throughout the text for simplicity of circuit analysis unless otherwise specified.

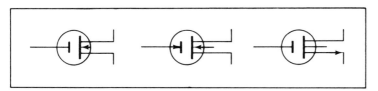

Figure 2-43 *Typical MOSFET Symbols*

problems

1. Given a single battery fixed bias circuit. If R_L is equal to $4\,k\Omega$ and the supply voltage V_{CC} is equal to 20 volts, what are the intercepts of the dc load line?

2. Use the circuit of problem 1, the required bias current in the base circuit is $100\,\mu A$. Determine the value of R_B.

3. The self bias circuit of Fig. 2-17 has a V_{CC} of 18 V, an R_L of $4.5\,k\Omega$ and $R_B = 200\,k\Omega$. Calculate I_B, I_C and I_E with a transistor having β equal to 50 and a leakage current of $10\,\mu A$.

4. The self bias circuit of Fig. 2-17 has a V_{CC} of 16 V an R_L of $3\,k\Omega$ and an $R_B = 150\,k\Omega$. Calculate I_B, I_C and I_E when a transistor having β equal to 40 and a leakage current of $20\,\mu A$.

5. The 2N35 transistor is used in the circuit shown.

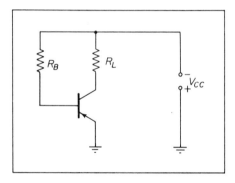

(a) If $V_{CC} = 16$ V, $R_L = 2\,k\Omega$ and $R_B = 200\,k\Omega$, determine the point of operation.

(b) If $V_{CC} = 18$ V, $R_L = 3\,k\Omega$ and $R_B = 300\,k\Omega$, determine the point of operation.

6. Given the circuit shown, determine the point of operation, the ac load line and S.

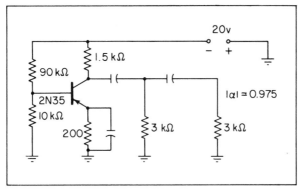

7. Given the circuit shown and the given measurements, determine: (a) R (b) the ac load line (c) S.

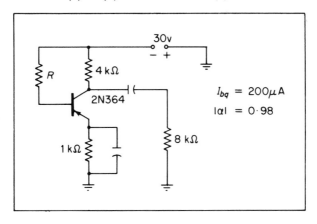

8. Given the circuit shown, determine the point of operation and the ac load line.

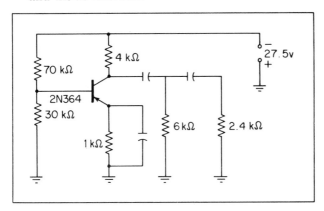

9. Given the circuit shown, determine the point of operation and the ac load line. What is the stability factor?

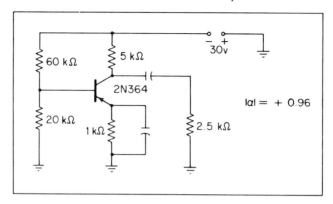

10. Given the circuit shown and the following measurements, determine: (a) the Q point, (b) the ac load line, (c) the stability factor.

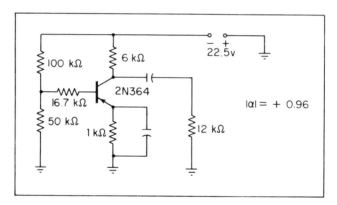

11. The 2N35 transistor is used in the circuit shown.

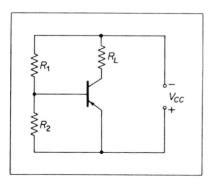

(a) If $V_{cc} = 16$ V, $R_L = 2$ kΩ, $R_1 = 180$ kΩ and $R_2 = 120$ kΩ determine the point of operation

(b) If $V_{cc} = 18$ V, $R_L = 3$ kΩ, $R_1 = 150$ kΩ and $R_2 = 50$ kΩ determine the point of operation

12. Find the stability factor for the given circuit.

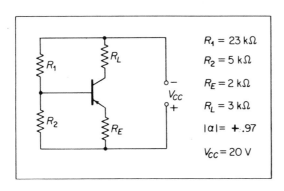

$R_1 = 23$ kΩ

$R_2 = 5$ kΩ

$R_E = 2$ kΩ

$R_L = 3$ kΩ

$|a| = +.97$

$V_{cc} = 20$ V

13. Find the stability factor for the given circuit.

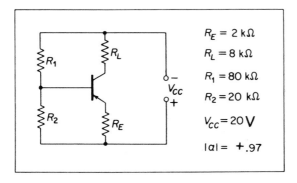

$R_E = 2$ kΩ

$R_L = 8$ kΩ

$R_1 = 80$ kΩ

$R_2 = 20$ kΩ

$V_{cc} = 20$ V

$|a| = +.97$

14. Find the stability factor for the given circuit.

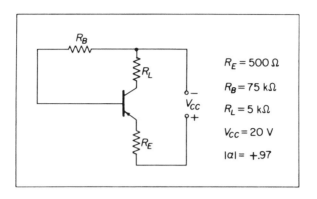

$R_E = 500\ \Omega$

$R_B = 75\ k\Omega$

$R_L = 5\ k\Omega$

$V_{CC} = 20\ V$

$|a| = +.97$

15. Find the stability factor for the given circuit.

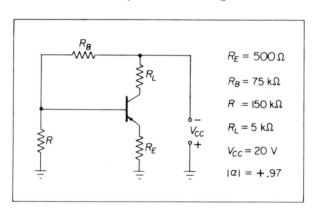

$R_E = 500\ \Omega$

$R_B = 75\ k\Omega$

$R = 150\ k\Omega$

$R_L = 5\ k\Omega$

$V_{CC} = 20\ V$

$|a| = +.97$

16. A 2N364 transistor is used in the circuit shown.

$$R_E = 500 \ \Omega$$
$$R_L = 6 \ k\Omega$$
$$V_{CC} = 30 \ V$$
$$\alpha = + .98$$
$$I_{C_q} = 3 \ mA$$

Find: (a) R_B and S
   (b) R_B and S when $R_E = 1 \ k\Omega$

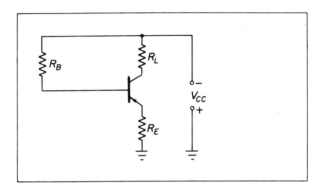

17. Find the stability factor for the given circuit.

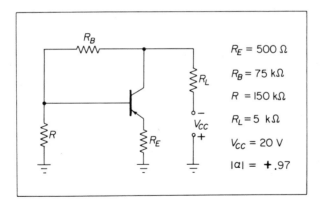

$R_E = 500 \ \Omega$

$R_B = 75 \ k\Omega$

$R = 150 \ k\Omega$

$R_L = 5 \ k\Omega$

$V_{CC} = 20 \ V$

$|\alpha| = + .97$

18. The following circuit is given with an operating point at $V_{ce} = 8$ V, $I_{c_q} = 4$ mA, $S = 12$. Find R_E, R_1 and R_2

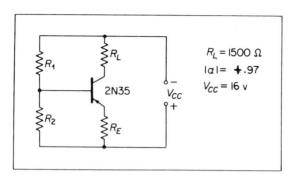

$R_L = 1500$ Ω
$|a| = +.97$
$V_{CC} = 16$ v

19. The following circuit is given with an operating point $V_{ce} = 15$ V, $I_{c_q} = 2.25$ mA, $S = 12$. Find R_E, R_1 and R_2.

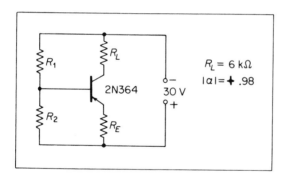

$R_L = 6$ kΩ
$|a| = +.98$

three

SMALL SIGNAL AMPLIFIERS

TRANSISTOR PARAMETERS

The behavior of any electrical circuit may be analyzed by mathematical as well as graphic procedures. Graphic techniques are usually utilized for analyzing large signal level stages or for the evaluation of the operating point of an amplifier.

The standard method for evaluating system performance is by means of small signal parameters. These parameters are usually supplied by the manufacturer or determined by test procedures.

An electrical equivalent circuit or model will be used to represent the transistor. Since the transistor is an active device and nonlinear, the relationship between voltage and current will be assumed linear over a limited range of their operating characteristics.

The elements of the equivalent circuit for the transistor can be developed from the terminal properties of the device. When analyzing a transistor amplifier, two pairs of terminals are considered—the input and output terminals. Consider a four-terminal network shown in Fig. 3-1.

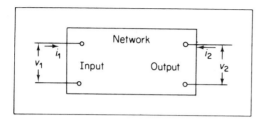

Figure 3-1 *General Network Showing Current and Voltage Directions*

The external conditions are measurable, namely, v_1, i_1, v_2 and i_2. Any two quantities may be arbitrarily chosen as the dependent quantities with the other two quantities becoming the independent variables. Each choice leads to the development of a different equivalent circuit. For example, let the voltages v_1 and v_2 be chosen as the dependent variables. The following mathematical statements can be written.

$$v_1 = f(i_1, i_2)$$
$$v_2 = f(i_1, i_2)$$

and the mathematical equations are:

$$v_1 = i_1 z_{11} + i_2 z_{12}$$
$$v_2 = i_1 z_{21} + i_2 z_{22}$$

These two equations completely describe the relationship that exists between the terminal voltages and currents. These equations apply regardless of the exact arrangement existent within the box labeled "network."

The following definitions are given for the "z or impedance parameters."

$$z_{11} = \frac{v_1}{i_1}\Big|_{i_2=0}$$ (Input impedance with the output terminals open circuited)

$$z_{12} = \frac{v_1}{i_2}\Big|_{i_1=0}$$ (reverse transfer impedance with the input terminals open)

$$z_{21} = \frac{v_2}{i_1}\Big|_{i_2=0}$$ (forward transfer impedance with the output terminals open)

$$z_{22} = \frac{v_2}{i_2}\Big|_{i_1=0}$$ (output impedance with the input terminals open)

Examination of the definitions shows that all of these parameters involve the ratio of a voltage to a current; therefore, they have the dimensions of impedance. Note that each one is defined only when one of the currents in the network is zero, thus, the name open circuit parameters.

Specifically, for a transistor used in the grounded base circuit configuration, the input current i_1 is I_e, the input voltage v_1 is v_e, the output current i_2 is i_c and the output voltage v_2 becomes v_c. The complete equivalent circuit of the transistor with both signal and load is illustrated in Fig. 3-2.

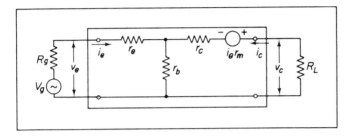

Figure 3-2   *Transistor Equivalent Circuit*

The resistive parameters are defined as follows:

r_e = lead in emitter resistance
r_b = lead in base resistance
r_c = lead in collector resistance
r_m = mutual resistance

The basic assumption in arriving at a transistor linear model or equivalent circuit is that the variations about the Q point are assumed small so that the transistor parameters may be considered linear. The model presented is one of a few possible models.

The following procedures regarding Kirchhoff's law equations will be followed in circuit analysis. The closed arrow denotes the direction of conventional current flow. The head of the arrow is positive and the tail is negative in the voltage equations.

Although these resistive parameters are no longer widely used for the specifications of transistors, they are useful in that they are readily understandable. Thus, the equations for the circuit of Fig. 3-3 are;

$$V_e = V_g - I_e R_g = I_e z_{11} + I_c z_{12}$$
$$V_c = -I_c R_L = I_e z_{21} + I_c z_{22}$$

Combining terms and simplifying yields

$$V_g = I_e(R_g + z_{11}) + I_c z_{12}$$
$$0 = I_e z_{21} + I_c(R_L + z_{22})$$

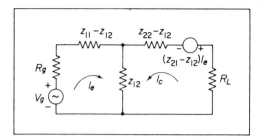

Figure 3-3   *General T Network for a Transistor*

The current amplification is defined as the ratio of output current to input current. Thus,

$$A_i = \frac{I_c}{I_e} = -\frac{z_{21}}{z_{22} + R_L}$$

The input impedance of the circuit is defined by the following relationship:

$$R_{in} = \frac{V_g}{I_e} - R_g$$

where:

$$I_e = \frac{V_g(z_{22} + R_L)}{(R_g + z_{11})(R_L + z_{22}) - z_{12}z_{21}}$$

Combining terms and simplifying yields:

$$R_{in} = z_{11} - \frac{z_{12}z_{21}}{z_{22} + R_L}$$

The voltage amplification of the device is given by:

$$A_v = \frac{V_c}{V_g} = \frac{-I_c R_L}{V_g}$$

The value of I_c is:

$$I_c = \frac{-V_g z_{21}}{(R_g + z_{11})(R_L + z_{22}) - z_{12}z_{21}}$$

Combining terms and simplifying yields:

$$A_v = \frac{z_{21}R_L}{(R_g + z_{11})(R_L + z_{22}) - z_{12}z_{21}}$$

or

$$A_v = \frac{-A_i R_L}{R_g + R_{in}}$$

The output impedance is given by:

$$R_o = z_{22} - \frac{z_{12}z_{21}}{R_g + z_{11}}$$

Note that the z parameters are measured values at the terminals of the transistor. The equivalent T circuit is a network concept that is used to represent the transistor. The relationship between the values r_e, r_b, r_c and r_m and the z parameters are given in Table 3-1 for the three possible transistor circuit configurations.

An illustrative problem will demonstrate the theory.

TABLE 3-1   *z PARAMETER RELATIONSHIPS*

	CB	CE	CC
z_{11}	$r_e + r_b$	$r_b + r_e$	$r_b + r_c$
z_{12}	r_b	r_e	$r_c - r_m$
z_{21}	$r_b + r_m$	$r_e - r_m$	r_c
z_{22}	$r_b + r_c$	$r_e + r_c - r_m$	$r_e + r_c - r_m$

sample problem

The following circuit has the given measurements:

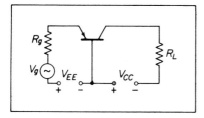

$$r_e = 25 \; \Omega \qquad r_c = 3 \; \text{Meg} \; \Omega \qquad R_g = 500 \; \Omega$$

$$r_b = 250 \; \Omega \qquad r_m = 2.94 \; \text{Meg} \; \Omega \qquad R_L = 10 \; \text{k}\Omega$$

Find: (a) A_i (b) R_{in} (c) A_v (d) R_o

Solution:

Step 1:   The circuit is a common base circuit. Calculate A_i:

$$A_i = -\frac{z_{21}}{z_{22} + R_L}$$

$$A_i = -\left(\frac{r_b + r_m}{r_c + r_b + R_L}\right)$$

$$A_i = -.98$$

Step 2:   Calculate R_{in}

$$R_{in} = z_{11} - \frac{z_{12} z_{21}}{R_L + z_{22}}$$

$$R_{in} = r_e + r_b - \frac{r_b(r_b + r_m)}{R_L + r_b + r_c}$$

$$R_{in} = 275 - 245 = 30 \; \Omega$$

Step 3:   Calculate A_v

$$A_v = \frac{-A_i R_L}{R_g + R_{in}}$$

$$A_v = \frac{+(.98)10^4}{530}$$

$$A_v = 18.5$$

Step 4:   Calculate R_0

$$R_o = z_{22} - \frac{z_{12} z_{21}}{R_g + z_{11}}$$

$$R_o = r_c + r_b - \frac{r_b(r_b + r_m)}{R_g + r_e + r_b}$$

$$R_o \cong 3 \times 10^6 - \frac{250(2.94 \times 10^6)}{775}$$

$$R_o \cong 2.05 \times 10^6 \; \text{ohms.}$$

HYBRID PARAMETERS

For simplicity and ease of analysis, the z parameters and equivalent T circuit were used. The measurement of these parameters is extremely difficult; consequently, manufacturers developed the *hybrid parameters*, which were simple and easy to measure. The choice previously made of the dependent and independent variables resulted in the open circuit equivalent T network parameters. If the voltage v_1 and the current i_2 are chosen as the dependent variables and the remaining two quantities are the independent variables, the resulting equations are:

$$v_1 = i_1 h_{11} + v_2 h_{12}$$
$$i_2 = i_1 h_{21} + v_2 h_{22}$$

The following definitions are given for the "h or hybrid parameters":

$$h_{11} = \frac{v_1}{i_1}\bigg|_{v_2=0} \quad \text{(input impedance with the output terminals short circuited)}$$

$$h_{12} = \frac{v_1}{v_2}\bigg|_{i_1=0} \quad \text{(reverse open circuit voltage amplification factor)}$$

$$h_{21} = \frac{i_2}{i_1}\bigg|_{v_2=0} \quad \text{(forward short circuit current amplification factor)}$$

$$h_{22} = \frac{i_2}{v_2}\bigg|_{v_1=0} \quad \text{(output admittance with input open circuited)}$$

Because of the mixture of both open and short circuit measurements, these values are called *hybrid parameters*.

To standardize transistor terminology, the IEEE (Institute of Electronic and Electrical Engineers) utilizes the following parameter symbols:

$$h_{i(\ )} = h_{11}$$
$$h_{r(\ )} = h_{12}$$
$$h_{f(\ )} = h_{21}$$
$$h_{o(\ )} = h_{22}$$

The additional subscripts, (e), (b) and (c), are used to identify the circuit configuration whether it is a common emitter, common base or common collector. The various symbols and their meaning are given in Table 3-2.

TABLE 3-2 *TRANSISTOR PARAMETER SYMBOLS*

	CB	CE	CC
h_{11}	h_{ib}	h_{ie}	h_{ic}
h_{12}	h_{rb}	h_{re}	h_{rc}
h_{21}	h_{fb}	h_{fe}	h_{fc}
h_{22}	h_{ob}	h_{oe}	h_{oc}

Figure 3-4 illustrates the three basic equivalent circuits using the recommended nomenclature.

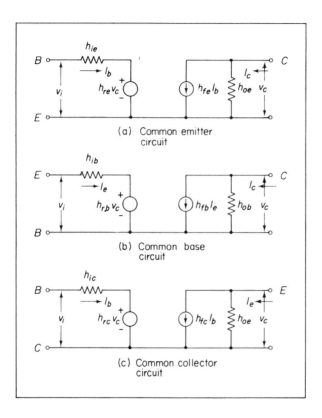

(a) Common emitter circuit

(b) Common base circuit

(c) Common collector circuit

Figure 3-4   *h Parameter Equivalent Circuit: (a) Common Emitter Circuit (b) Common Base Circuit (c) Common Collector Circuit*

The analysis of the grounded emitter circuit will be performed first and the equations derived can also be made applicable to the grounded base and grounded collector circuits.

analysis   Assume a generator with its internal impedance connected to each basic circuit. Consider the circuit shown in Fig. 3-5.

The operating equations for the given circuit are:

$$V_g = (h_{ie} + R_g)I_b + h_{re}v_c$$
$$I_c = h_{fe}I_b + h_{oe}v_c$$
$$v_c = -I_c R_L$$

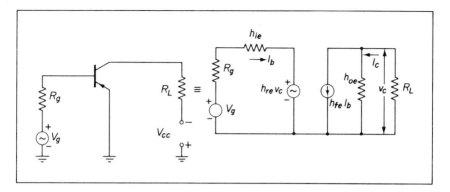

Figure 3-5 *Common Emitter Circuit*

The two basic equations are:

$$V_g = (h_{ie} + R_g)I_b + h_{re}v_c$$

$$0 = h_{fe}I_b + \left(h_{oe} + \frac{1}{R_L}\right)v_c$$

Substituting for v_c in the second equation yields:

$$0 = h_{fe}I_b + \left(h_{oe} + \frac{1}{R_L}\right)(-I_cR_L)$$

The current amplification is defined by the ratio of the output current to the input current. Thus,

$$A_{ie} = \frac{I_c}{I_b}$$

$$A_{ie} = \frac{h_{fe}}{1 + h_{oe}R_L}$$

The input resistance of the transistor is determined by

$$R_{ie} = \frac{V_g}{I_b} - R_g$$

Using the equation,

$$V_g = (R_g + h_{ie})I_b + h_{re}(-I_cR_L)$$

Dividing both sides by I_b and subtracting R_g from both sides yields the resultant equation for R_{ie}. Thus,

$$R_{ie} = h_{ie} - h_{re}A_{ie}R_L$$

The voltage amplification of the circuit is given by:

$$A_v = \frac{V_o}{V_g}$$

The output voltage, V_o, is equal to v_c. Thus solving the given equations for v_c yields:

$$v_c = \frac{-h_{fe}V_g}{(h_{ie} + R_g)\left(h_{oe} + \frac{1}{R_L}\right) - h_{re}h_{fe}}$$

Further simplification of the equation yields

$$A_{ve} = \frac{-A_{ie}R_L}{R_g + R_{ie}}$$

The output resistance R_{oe} is evaluated by removing the input generator and replacing the generator by its internal impedance. Thus, the resultant equations are:

$$0 = (R_g + h_{ie})I_b + h_{re}v_c$$
$$I_c = h_{fe}I_b + h_{oe}v_c$$

Solving these equations for v_c results in:

$$v_c = \frac{I_c(R_g + h_{ie})}{(R_g + h_{ie})h_{oe} - h_{re}h_{fe}}$$

The value of R_o is determined by the ratio of v_c to I_c. Thus,

$$R_{oe} = \frac{R_g + h_{ie}}{(R_g + h_{ie})h_{oe} - h_{re}h_{fe}}$$

The output conductance is the reciprocal of R_o. Thus,

$$G_{oe} = h_{oe} - \frac{h_{re}h_{fe}}{R_g + h_{ie}}$$

The actual power gain of the circuit is given by:

$$P.G. = A_{ie} \cdot A_{ve}$$

or

$$P.G. = \frac{A_{ie}^2 R_L}{R_g + R_{ie}}$$

The equations for the input resistance, output resistance, current and voltage amplification and the power gain that were derived for the grounded emitter circuit are also applicable to the common base and common collector circuits providing the appropriate values of the "h parameters" are used. Consequently, the "h parameters" can readily be used to evaluate system performance. The following summary of the required operating equations for the different circuit configurations is given for the convenience of the student.

$$A_{i_{(\,)}} = \frac{h_{f_{(\,)}}}{1 + h_{o_{(\,)}}R_L}$$

$$R_{i_{(\,)}} = h_{i_{(\,)}} - h_{r_{(\,)}}A_{i_{(\,)}}R_L$$

$$A_{v_{(\,)}} = \frac{-A_{i_{(\,)}}R_L}{R_g + R_{i_{(\,)}}}$$

$$P.G._{(\,)} = A_{i_{(\,)}} \cdot A_{v_{(\,)}}$$

$$G_{o_{(\,)}} = h_{o_{(\,)}} - \frac{h_{r_{(\,)}}h_{f_{(\,)}}}{R_g + h_{i_{(\,)}}}$$

The parentheses permit evaluation of the circuit performance depending on the type of circuit configuration. In many cases, it may be necessary to interrelate the common base, common emitter and common collector circuit configurations. The required "h parameter" conversions are specified in Table 3-3.

An illustrative problem will demonstrate the theory.

sample problem

A common emitter transistor circuit has the following given data:

$$h_{ie} = 1.5 \text{ k}\Omega \quad h_{re} = 7 \times 10^{-4} \quad\quad R_g = 200 \text{ ohms.}$$
$$h_{fe} = 49 \quad\quad h_{oe} = 30 \times 10^{-6} \text{ mhos} \quad R_L = 3 \text{ k}\Omega$$

Find: (a) A_i, R_i, A_v, $P.G.$ and R_o for a *CE* circuit.
     (b) R_i, A_i, A_v, $P.G.$ and R_o for a *CB* circuit.
     (c) A_i, R_i, A_v, $P.G.$ and R_o for a *CC* circuit.

Solution:

Step 1: Calculate A_i for a *CE* circuit

$$A_i = \frac{h_{fe}}{1_{oe} + h_{oe}R_L}$$

$$A_i = \frac{49}{1 + .09}$$

$$A_i = 45$$

Step 2:   Calculate R_i

$R_i = h_{ie} - A_{i_e}h_{re}R_L$

$R_i = 1500 - 45 \times 7 \times 3 \times 10^{-1}$

$R_i = 1405.5$ ohms

Step 3:   Calculate A_v

$A_v = \dfrac{-A_{ie}R_L}{R_g + R_{ie}}$

$A_v = \dfrac{-45 \times 3 \times 10^3}{200 + 1405.5}$

$A_v = -84$

Step 4:   Calculate *P.G.*

$P.G. = A_i \times A_v$

$P.G. = 45 \times 84$

$P.G. = 3780$

Step 5:   Calculate R_o

$G_o = h_{oe} - \dfrac{h_{re}h_{fe}}{R_g + h_{ie}}$

$G_o = 9.8 \times 10^{-6}$ mhos

$R_o = \dfrac{1}{G_o} = 102.5$ kΩ

Step 6:   The solution of the *CB* circuit requires conversion of the *CE* parameters to *CB* parameters. Refer to Table 3-3. Thus,

$h_{ib} = 30$ ohms $h_{rb} = 2 \times 10^{-4}$

$h_{fb} = -.98$ $h_{ob} = .6 \times 10^{-6}$ mhos.

Step 7:   Substitution of the numbers into the required formulas yields the following results:

$A_{i_b} = -.978$

$R_{i_b} = 30.6$ ohms

$A_{v_b} = 12.72$

$P.G. = 12.44$

$R_o = 688$ kΩ

Step 8: The solution of the *CC* circuit requires conversion of the *CE* parameters to the *CC* parameters. Refer to Table 3-3. Thus,

$h_{ic} = 1.5\text{ k}\Omega$ $\qquad h_{rc} = .9993$

$h_{fc} = -50$ $\qquad h_{oc} = 30 \times 10^{-6}$ mhos.

TABLE 3-3 *APPROXIMATE CONVERSION FORMULAS*

Parameter	Common emitter	Common base	Common collector
h_{ie}		$\dfrac{h_{ib}}{1 + h_{fb}}$	h_{ic}
h_{re}		$\dfrac{h_{ob}h_{ib}}{1 + h_{fb}} - h_{rb}$	$1 - h_{rc}$
h_{fe}		$\dfrac{-h_{fb}}{1 + h_{fb}}$	$-(1 + h_{fc})$
h_{oe}		$\dfrac{h_{ob}}{1 + h_{fb}}$	h_{oc}
h_{ib}	$\dfrac{h_{ie}}{1 + h_{fe}}$		$-\dfrac{h_{ic}}{h_{fc}}$
h_{rb}	$\dfrac{h_{ie}h_{oe}}{1 + h_{fe}} - h_{re}$		$h_{rc} - 1 - \dfrac{h_{ic}h_{oc}}{h_{fc}}$
h_{fb}	$\dfrac{-h_{fe}}{1 + h_{fe}}$		$\dfrac{(1 + h_{fc})}{h_{fc}}$
h_{ob}	$\dfrac{h_{oe}}{1 + h_{fe}}$		$\dfrac{-h_{oc}}{h_{fc}}$
h_{ic}	h_{ie}	$\dfrac{h_{ib}}{1 + h_{fb}}$	
h_{rc}	$1 - h_{re}$	≈ 1	
h_{fc}	$-(1 + h_{fe})$	$\dfrac{-1}{1 + h_{fb}}$	
h_{oc}	h_{oe}	$\dfrac{h_{ob}}{1 + h_{fb}}$	

Step 9: Substitution of numbers into the required formulas yields the following results:

$A_{i_c} = -45.9$ $\qquad A_{v_c} = .989$

$R_{i_c} = 139$ kilohms $\quad P.G. = 45.4$

$R_{o_c} = 34$ ohms

The three circuits are compared in a summarized table given below. Thus,

	CB	*CE*	*CC*
A_i	$-.978$	45	-45.9
R_i	30.6 ohms	1405.5 ohms	139 kilohms
A_v	12.72	-84	.989
P.G.	12.44	3780	45.4
R_o	688 kilohms	102 kilohms	34 ohms

ADMITTANCE PARAMETERS

Other parameters that are extremely useful in the high frequency region are called the "y or admittance parameters." Let the voltages v_1 and v_2 be selected as the independent variables and the currents i_1 and i_2 become the dependent variables. The following mathematical relationships can be written.

$$i_1 = f(v_1, v_2)$$
$$i_2 = f(v_1, v_2)$$

The basic mathematical equations are:

$$i_1 = y_{11}v_1 + y_{12}v_2$$
$$i_2 = y_{21}v_1 + y_{22}v_2$$

These two equations completely describe the relationship that exists between terminal currents and voltages. The following definitions are given for the "y or admittance parameters."

$$y_{11} = \frac{i_1}{v_1}\bigg|_{v_2=0} \quad \text{(input admittance with the output short circuited)}$$

$$y_{12} = \frac{i_1}{v_2}\bigg|_{v_1=0} \quad \text{(forward transfer admittance with input short circuited)}$$

$$y_{21} = \frac{i_2}{v_1}\bigg|_{v_2=0} \quad \text{(reverse transfer admittance with output short circuited)}$$

$$y_{22} = \frac{i_2}{v_2}\bigg|_{v_1=0} \quad \text{(output admittance with the input short circuited)}$$

The equivalent circuit representing the transistor model incorporating the y parameters is shown in Fig. 3-6.

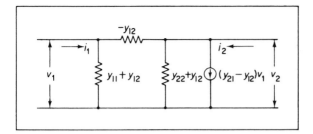

Figure 3-6 *Equivalent Circuit Using y Parameters*

It is evident that since some equivalent circuits have been developed, a method must exist for converting from one set of parameters to another set of required parameters. Table 3-4 shows the interrelationship of the developed network parameters.

TABLE 3-4 *MATRIX INTERRELATIONSHIP OF NETWORK PARAMETERS**

From → To ↓	z		y		h	
z	z_{11}	z_{12}	$\dfrac{y_{22}}{\Delta y}$	$\dfrac{-y_{12}}{\Delta y}$	$\dfrac{\Delta h}{h_{22}}$	$\dfrac{h_{12}}{h_{22}}$
	z_{21}	z_{22}	$\dfrac{-y_{21}}{\Delta y}$	$\dfrac{y_{11}}{\Delta y}$	$\dfrac{-h_{21}}{h_{22}}$	$\dfrac{1}{h_{22}}$
y	$\dfrac{z_{22}}{\Delta z}$	$\dfrac{-z_{12}}{\Delta z}$	y_{11}	y_{12}	$\dfrac{1}{h_{11}}$	$\dfrac{-h_{12}}{h_{11}}$
	$\dfrac{-z_{21}}{\Delta z}$	$\dfrac{z_{11}}{\Delta z}$	y_{21}	y_{22}	$\dfrac{h_{23}}{h_{11}}$	$\dfrac{\Delta h}{h_{11}}$
h	$\dfrac{\Delta z}{z_{22}}$	$\dfrac{z_{12}}{z_{22}}$	$\dfrac{1}{y_{11}}$	$\dfrac{-y_{12}}{y_{11}}$	h_{11}	h_{12}
	$\dfrac{-z_{21}}{z_{22}}$	$\dfrac{1}{z_{22}}$	$\dfrac{y_{21}}{y_{11}}$	$\dfrac{\Delta y}{y_{11}}$	h_{21}	h_{22}

$\Delta z = z_{11}z_{22} - z_{12}z_{21}$

$\Delta y = y_{11}y_{22} - y_{12}y_{21}$

$\Delta h = h_{11}h_{22} - h_{12}h_{21}$

*B. Zeines: "Introduction to Network Analysis," Prentice-Hall, 1967, p. 20.

JFET EQUIVALENT CIRCUIT

An equivalent circuit utilizes the typical *y* parameters as shown in Fig. 3-7.

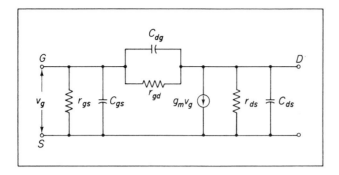

Figure 3-7 *Equivalent Circuit of the JFET*

The typical values of the equivalent circuit parameters are:

$G_m \cong 500$ to $20{,}000$ μmhos

$r_d \cong 2$ to 100 kΩ $\qquad\qquad C_{ds} \cong 0.1$ to 1 pF

$r_{gs} \geqq 10^9$ Ω $\qquad\qquad\quad C_{gs} \cong 1$ to 10 pF

$r_{gd} \geqq 10^{14}$ $\qquad\qquad\qquad C_{gd} \cong 1$ to 10 pF.

The various FET parameters are defined as follows. The dynamic drain resistance of the FET is:

$$\left.\frac{\Delta v_d}{\Delta i_d}\right|_{v_{gs}=k} = r_d(\text{dynamic drain resistance})$$

The amplification factor of the FET is given by:

$$\left.\frac{\Delta v_d}{\Delta v_{gs}}\right|_{i_d=k} = u \text{ (amplification factor)}$$

The transconductance of the FET is given by:

$$\left.\frac{\Delta i_d}{\Delta v_g}\right|_{v_d=k} = G_m \text{ (FET transconductance)}$$

For small signal linear operation, the interrelationships of the FET parameters are:

$$\frac{u}{r_d} = G_m$$

The family of characteristic curves for the FET represents in a graphical way the functional relationship of the drain current to the gate and drain voltages respectively. Thus,

$$i_d = f(v_g, v_d)$$

Assuming small signal linear operation, the drain current can be expressed mathematically as:

$$i_d = \frac{v_d + uv_g}{r_d}$$

The resultant small signal model is shown in Fig. 3-8.

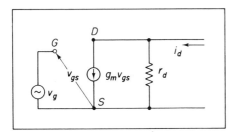

Figure 3-8 *Simplified FET Equivalent Circuit*

analysis The field effect transistor (FET) is a voltage operated device. Consider the circuit shown in Fig. 3-9.

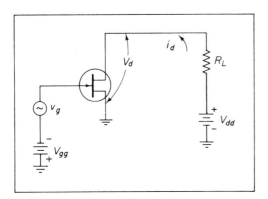

Figure 3-9 *Common Source Amplifier*

If the signal voltage is made negative, current flow decreases in the output circuit resulting in a voltage drop across R_L, increasing the drain voltage. Consequently, the amplifier has a 180 degree phase shift between input and output signal voltages.

The exact equivalent circuit is shown in Fig. 3-10.

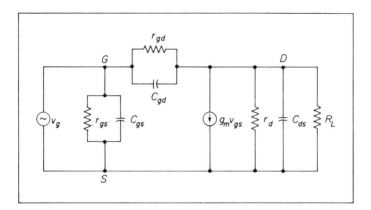

Figure 3-10   *Exact Equivalent Circuit*

Insertion of typical values yields the simplified equivalent circuit shown in Fig. 3-11.

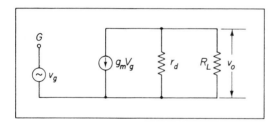

Figure 3-11   *Simplified Equivalent Circuit*

The voltage amplification of the circuit is determined by the ratio of the output voltage to the input voltage. The resultant mathematical expression is:

$$A_v = \frac{V_o}{V_g} = -G_m R_e$$

where:

$$R_e = \frac{r_d R_L}{r_d + R_L}$$

problems

1. Determine the z parameters for the given circuit.
2. Determine the h parameters for the given circuit.
3. Determine the y parameters for the given circuit.
   Circuits for problems 1, 2 and 3.

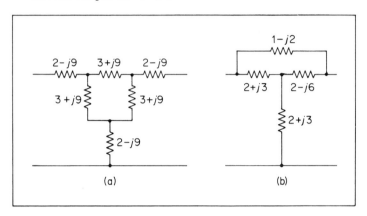

(a)   (b)

4. A given transistor has the following measurements specified:

   $h_{ib} = 30\ \Omega$ $\qquad\qquad$ $h_{rb} = 4 \times 10^{-4}$

   $h_{ob} = .56 \times 10^{-6}$ mhos $\qquad$ $h_{fe} = 35$

   Calculate the common emitter and common collector circuit parameters.

5. A transistor is used in a common emitter circuit. The specifications are:

   $h_{ib} = 25\ \Omega$ $\qquad$ $h_{rb} = 6.3 \times 10^{-4}$ $\quad$ $R_g = 250\ \Omega$

   $h_{ob} = .62 \times 10^{-6}$ $\quad$ $h_{fe} = 50$ $\qquad\qquad$ $R_L = 4\ \text{k}\Omega$

   Find: $A_i,\ R_i,\ A_v,\ P.G.,\ R_o$

6. Given the circuit shown, determine the h parameters.

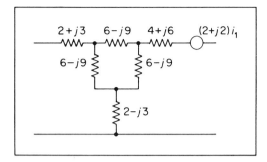

7. For the circuit of problem 6, determine the y parameters.

8. Given the circuit shown in the figure, determine the y parameters.

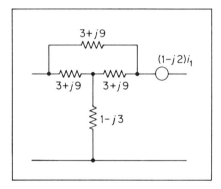

9. For the circuit of problem 8, determine the h parameters.

10. Given the circuit shown, determine the y parameters.

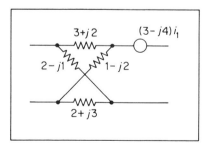

11. For the circuit of problem 10, determine the h paramaters.

four

TRANSISTORIZED VOLTAGE AMPLIFIERS

INTRODUCTION

The mathematical model for semiconductors can be used to determine the behavior and performance of small signal amplifiers.

In many practical systems, small signals must be tremendously amplified with a minimum amount of distortion to produce an output. The performance of an amplifier can be evaluated either in terms of (a) purpose (voltage, current or power amplification) or (b) frequency capabilities.

95

The classification of an amplifier can be determined from the mode of operation.

A Class A amplifier is one in which the angle of collector or drain current conduction is 360 degrees. This means that there is a continuous flow of output current throughout the entire cycle of input signal (see Fig. 4-1).

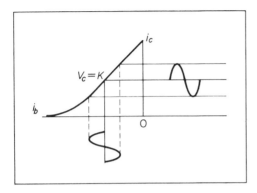

Figure 4-1   *Class A Operation*

A Class B amplifier is one in which the angle of collector or drain current conduction is 180 degrees. The amplifier is usually operated at cutoff and operates only when the signal input is above cutoff, as shown in Fig. 4-2.

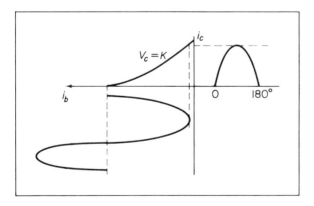

Figure 4-2   *Class B Operation*

A Class C amplifier is one in which the angle of collector or drain current conduction is less than 180 degrees. The "magic number" in industry is 140–150 degrees. The output current is a pulse, shown in Fig. 4-3, that contains a large number of harmonics. Thus, circuits of this type are used for high power output where the harmonic components can be removed without disturbing the fundamental component of the current pulse.

An amplifier can operate in the range between two classes. For example, a Class AB amplifier has an angle of current conduction between 180 and 360 degrees. On the other hand, a Class BC amplifier does not exist.

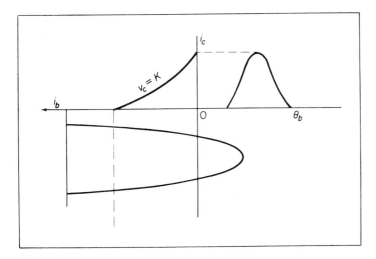

Figure 4-3 *Class C Amplifier*

In general, amplifiers are classified as either small signal or power amplifiers. The purpose of a small signal amplifier is to faithfully reproduce small signal inputs without distortion. The discussion in this Chapter will be restricted to Class A small signal operation of the amplifier.

COMMON EMITTER RC COUPLED AMPLIFIER

The interstage method of coupling is used extensively in describing the method of transferring energy from one stage to the next. The basic circuit of a resistance coupled (RC) small signal amplifier is shown in Fig. 4-4.

analysis A transistorized amplifier usually is evaluated by the amount of current amplification it has. The current amplification may be greater at some frequencies than others. The response characteristic is used to designate the amplification over a band of frequencies as shown in Fig. 4-5.

The range of frequencies shown extends throughout and beyond the audio frequency band, which starts approximately at 20 Hz and ends about 20 kHz. The dotted lines on the curve trisect the relative frequency ranges into low frequency, midfrequency and high frequency. The current amplification is maximum at the midfrequency range.

The boundaries for each range classification are not well defined. In

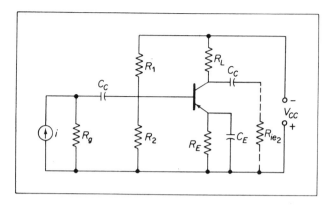

Figure 4-4   *RC Coupled Amplifier*

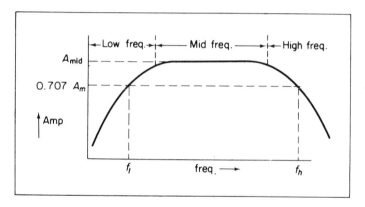

Figure 4-5   *Amplification vs Frequency for an RC Coupled Amplifier*

the application of small signals to audio frequency amplifiers, the limits of the response characteristic are defined as those frequencies where the current amplification response is equal to 0.707 of the midfrequency response. These are shown simply as f_1 and f_2 in Fig. 4-5, and are designated as the cutoff frequencies of the system.

In most practical cases, it is customary to determine the approximate values of amplification rather than the exact values.

The following assumptions will be made before proceeding with the analysis.

1. Since $\dfrac{1}{h_{oe}} \gg R_L$, then h_{oe} can be neglected. This results in $I_c \cong h_{fe}I_b$.

2. The reverse voltage transfer generator is usually small and can readily be neglected ($h_{re} \cong 0$).

The midfrequency region is designated as that range of frequencies in which the external capacitances are considered negligible. The simplified equivalent circuit is shown in Fig. 4-6. The midfrequency current amplification is determined by the ratio of the output current to the input current.

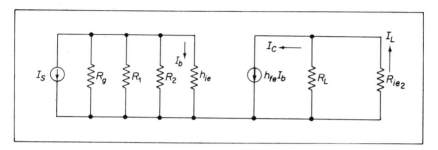

Figure 4-6 *Simplified Midfrequency Equivalent Circuit*

The basic operating equations are:

$$I_b = I_s \frac{R_s}{R_s + h_{ie}} \quad \text{where } R_s = \frac{1.}{\dfrac{1}{R_g} + \dfrac{1}{R_1} + \dfrac{1}{R_2}}$$

and

$$I_c = h_{fe} I_b \qquad I_L = \frac{I_c R_L}{R_L + R_{ie_2}}$$

Then

$$A_{i_{\text{mid}}} = \frac{I_L}{I_s} = \frac{h_{fe} R_s R_L}{(R_s + h_{ie})(R_L + R_{ie_2})}$$

The parallel combination of R_1 and R_2 usually results in a comparatively large resistor and may be neglected in the resultant equivalent circuit. Consequently, the value of R_s becomes equal to R_g; however, the R_s notation will be used.

LOW FREQUENCY ANALYSIS

Assume the emitter circuit only is to be analyzed. Refer to Fig. 4-7. The input circuit equation is:

$$v_b = I_b h_{ie} + (I_b + h_{fe} I_b) Z_E$$
or
$$v_b = I_b [h_{ie} + (1 + h_{fe}) Z_E]$$

which corresponds to the circuit shown in Fig. 4-8.

The equivalent low frequency circuit including the emitter circuit and neglecting the coupling capacitor is shown in Fig. 4-9.

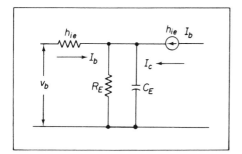

Figure 4-7   *Emitter Circuit Analysis*

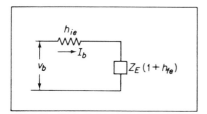

Figure 4-8

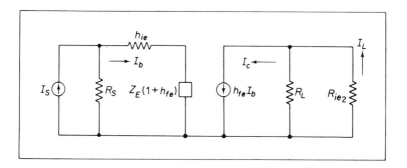

Figure 4-9   *Low Frequency Simplified Circuit*

analysis

$$A_{i_{\text{low}}} = \frac{I_L}{I_b} \frac{I_b}{I_s}$$

$$\frac{I_L}{I_s} = \frac{h_{fe} R_s}{R_s + h_{ie} + Z_E(1 + h_{fe})} \frac{R_L}{(R_L + R_{ie_2})}$$

$$A_{i_{\text{low}}} = \frac{A_{i_{\text{mid}}}}{1 + \dfrac{Z_E(1 + h_{fe})}{R_s + h_{ie}}}$$

Note that

$$Z_E = \frac{R_E}{1 + j\omega R_E C_E}$$

Substituting and simplifying yields the equation:

$$A_{i_{low}} = \frac{A_{i_{mid}}(1 + j\omega R_E C_E)}{1 + \dfrac{R_E(1 + h_{fe})}{R_s + h_{ie}} + j\omega R_E C_E}$$

and

$$A_{i_{low}} = \frac{A_{i_{mid}}}{K_0}\left(\frac{1 + j\dfrac{\omega}{\omega_1}}{1 + j\dfrac{\omega}{\omega_2}}\right)$$

where:

$$\omega_1 = \frac{1}{R_E C_E} \qquad K_0 = 1 + \frac{R_E(1 + h_{fe})}{R_s + h_{ie}}$$

$$\omega_2 = K_0 \omega_1$$

The response of the low frequency system is shown in Fig. 4-10.

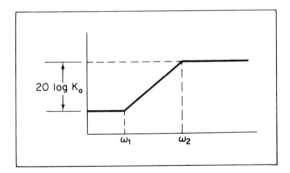

Figure 4-10 *Response Due to the Emitter Circuit*

The coupling capacitor is inserted into the circuit. The resulting equivalent circuit is shown in Fig. 4-11.

analysis

$$A_{i_{low}} = \frac{I_b}{I_s}\frac{I_L}{I_b}$$

$$A_{i_{low}} = \frac{R_s}{R_s + h_{ie} + Z_E(1 + h_{fe})}\frac{h_{fe}R_L}{R_{ie_2} + R_L + \dfrac{1}{j\omega C_c}}$$

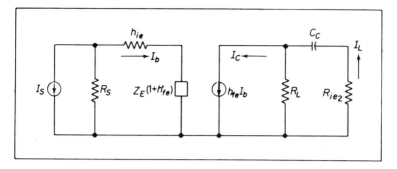

Figure 4-11   *Complete Simplified Low Frequency Model*

Substituting and simplifying yields the resultant equation:

$$A_{i_{low}} = \left(\frac{A_{i_{mid}}}{K_0}\right)\frac{\left(1 + j\frac{\omega}{\omega_1}\right)}{\left(1 + j\frac{\omega}{\omega_2}\right)}\frac{\left(j\frac{\omega}{\omega_c}\right)}{\left(1 + j\frac{\omega}{\omega_c}\right)}$$

where: $\omega_c = \dfrac{1}{(R_{ie_2} + R_L)C_c}$

$K_0 = 1 + \dfrac{R_E(1 + h_{fe})}{R_s + h_{ie}}$

$\omega_1 = \dfrac{1}{R_E C_E}$

$\omega_2 = K_0\omega_1$

The overall resultant low frequency response is shown in Fig. 4-12. Note that there are two break frequencies, namely, ω_c and ω_1.

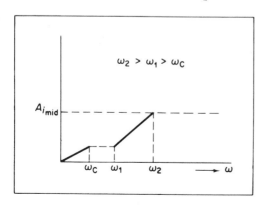

Figure 4-12   *Complete Low Frequency Response*

The half power point frequency at the low end is generally controlled by the emitter circuit providing $K_0 \geqq 5$. Consequently, if $\omega > \begin{cases} \omega_1 \\ \omega_c \end{cases}$

then $\qquad A_{i_{low}} = \dfrac{A_{i_{mid}}}{K_0} \left[\dfrac{j\dfrac{\omega}{\omega_1}}{1 + j\dfrac{\omega}{\omega_2}} \right]$

and simplifying, the resultant equation is:

$$A_{i_{low}} = A_{i_{mid}} \left[\dfrac{1}{1 - j\left(\dfrac{\omega_2}{\omega}\right)} \right]$$

where ω_2 is the half power point frequency in the lower audio frequency region. This frequency is generally designated by f_l. Thus,

$$A_{i_{low}} = A_{i_{mid}} \left[\dfrac{1}{1 - j\dfrac{f_l}{f}} \right]$$

sample problem

The following circuit is given:

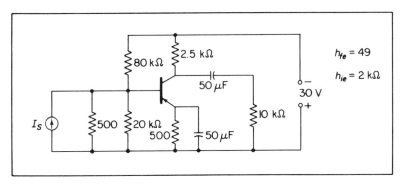

Find the low frequency response and sketch the resultant value.

Solution:

Step 1: Calculate $A_{i_{mid}}$

$$A_{i_{mid}} = \dfrac{h_{fe}R_sR_L}{(R_s + h_{ie})(R_L + R_{ie_2})}$$

$$A_{i_{mid}} = \dfrac{49 \times 500 \times 10^4 \times .25}{(500 + 2000)(2.5 + 10)10^3}$$

$$A_{i_{mid}} = 1.96$$

Step 2:   Calculate K_0

$$K_0 = 1 + \frac{R_E(1 + h_{fe})}{R_s + h_{ie}}$$

$$K_0 = 1 + 10 = 11$$

Step 3:   Calculate f_1

$$f_1 = \frac{1}{2\pi R_E C_E} = \frac{1}{2\pi (500)(50)10^{-6}}$$

$$f_1 = 6.36 \text{ Hz}$$

Step 4:   Calculate f_c

$$f_c = \frac{1}{2\pi (R_L + R_{ie_2})C_c} = \frac{1}{2\pi (2.5 + 10)10^3 \times 50 \times 10^{-6}}$$

$$f_c = 0.256 \text{ Hz}$$

Step 5:   Calculate f_2

$$f_2 = K_0 f_1$$

$$f_2 = 11(6.36)$$

$$f_2 = 69.96 \text{ Hz}$$

Step 6:   Construct the low frequency response:

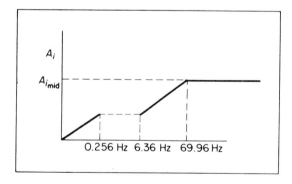

It is evident that the cutoff frequency at the low frequency end is approximately 70 Hz. Frequencies below this value do not affect the response. Note that the emitter circuit controls the low frequency response.

MILLER EFFECT

In any amplifier, a defined input impedance is present between base and emitter due to the circuit elements on the collector side. In the transistor manual, a capacitance from base to ground is usually listed. This value is determined without load parameters. Resistance in the collector circuit will increase the capacitance between base and ground. As the load resistance varies, the amount of capacitance reflected to the base circuit varies accordingly, and this phenomenon is known as the *Miller effect.* Consequently, the amount of capacitance reflected to the input side is dependent on the voltage amplification of the stage.

The presence of the input capacitance will affect the operation of the circuit particularly at the high frequencies. The capacitance at the input of the circuit exceeds any of the individual transistor capacitances since the base to collector capacitor is multiplied by the voltage amplification of the stage and then placed at the input side. If the voltage amplification term contains a real plus an imaginary component, then the equivalent circuit of the input impedance will consist of a resistor in parallel with a capacitor. Should the load contain an inductive element, the reflected value of the resistor is negative. A negative resistance denotes positive feedback or an oscillatory condition. With a purely resistive load, this problem will not exist.

The high frequency equivalent circuit incorporates the capacitors due to the base to emitter depletion region and the collector to base depletion region, and are shown in Fig. 4-13.

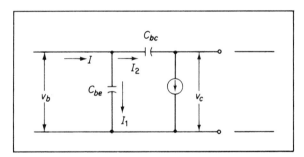

Figure 4-13 *High Frequency Equivalent Circuit*

analysis The current I is given by:

$$I = I_1 + I_2$$

The current I_1 is defined by:

$$I_1 = v_b(j\omega C_{be})$$

The current I_2 is given by:

$$I_2 = (v_b - v_c)j\omega C_{bc}$$

Note that $v_c = A_v v_b$; therefore, substituting for v_c results in:

$$I_2 = (v_b - A_v v_b) j \omega C_{bc}$$

The total current I is given by:

$$I = v_b [j \omega C_{be} + (1 - A_v) j \omega C_{bc}]$$

The input admittance is the ratio of the input current to the input voltage. Thus,

$$\frac{I}{v_b} = Y_{in} = j \omega \left[C_{be} + (1 - A_v) C_{bc} \right]$$

The coefficient of ω is equal to the input capacitance. Thus,

$$C_{in} = C_{be} + (1 - A_v) C_{bc} = C_T$$

The resultant equivalent circuit is shown in Fig. 4-14.

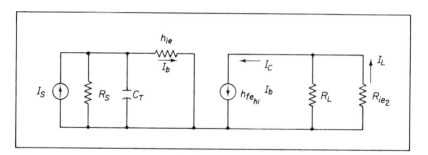

Figure 4-14   *Simplified Equivalent Circuit*

The current I_b is given by:

$$I_b = \frac{I_s \left(\dfrac{R_s}{1 + j \omega R_s C_T} \right)}{h_{ie} + \dfrac{R_s}{1 + j \omega C_T R_s}}$$

The equation can be simplified to:

$$I_b = \frac{I_s R_s}{(R_s + h_{ie})} \left[\frac{1}{1 + j \omega \left(\dfrac{R_s h_{ie}}{R_s + h_{ie}} \right) C_T} \right]$$

Since $I_c = h_{fe_{hi}}I_b$ and $I_L = \dfrac{I_c R_L}{R_L + R_{ie_2}}$, then the ratio of the output current to the input current is given by:

$$\frac{I_L}{I_s} = \frac{h_{fe_{hi}}R_L \ R_s}{(R_L + R_{ie_2})(R_s + h_{ie})\left(1 + j\dfrac{f}{f_h}\right)}$$

where:

$$f_h = \frac{1}{2\pi\left(\dfrac{R_s h_{ie}}{R_s + h_{ie}}\right)C_T}$$

Note that the relative response is given by the ratio of the current amplification at the high frequencies to the current amplification at the mid-frequencies. Thus,

$$\rho = \frac{A_{i_{hi}}}{A_{i_{mid}}}$$

In the development of the various sets of network parameters, it was noted that each value was a function of frequency. The short circuit current amplification factor decreases as the frequency increases. Consequently, the relation between the network parameter h_{fb} and frequency can be expressed mathematically as:

$$h_{fb_{hi}} = \frac{|h_{fb_{mid}}|}{1 + j\dfrac{f}{f_a}}$$

where

$|h_{fb_{mid}}|$ = value of the midfrequency short circuit current amplification factor for a grounded base circuit.

f_a = frequency where the short circuit current amplification factor drops to 0.707 of its midfrequency value for a grounded base amplifier.

The short circuit current amplification factor for a grounded emitter circuit is given by the equation:

$$h_{fe_{mid}} = \frac{|h_{fb_{mid}}|}{1 - |h_{fb_{mid}}|}$$

Assuming the amplifier is operating in the high frequency region, the short circuit current amplification factor for a grounded emitter is given by:

$$h_{fe_{hi}} = \frac{|h_{fb_{hi}}|}{1 - |h_{fb_{hi}}|}$$

$$h_{fe_{hi}} = \frac{\dfrac{|h_{fb_{mid}}|}{1 + j(f/f_\alpha)}}{1 - \dfrac{|h_{fb_{mid}}|}{1 + j(f/f_\alpha)}} = \frac{|h_{fb_{mid}}|}{1 - |h_{fb_{mid}}| + j\dfrac{f}{f_\alpha}}$$

Simplifying yields the resultant equation,

$$h_{fe_{hi}} = \frac{|h_{fe_{mid}}|}{1 + j\dfrac{f}{f_\alpha(1 - |h_{fb_{mid}}|)}}$$

Let $f_\beta = f_\alpha(1 - |h_{fb_{mid}}|)$

The overall relative response of the circuit can be expressed mathematically as:

$$\rho = \frac{1}{\left(1 + j\dfrac{f}{f_h}\right)\left(1 + j\dfrac{f}{f_\beta}\right)}$$

An illustrative problem will demonstrate the theory.

sample problem

An RC coupled circuit is shown.

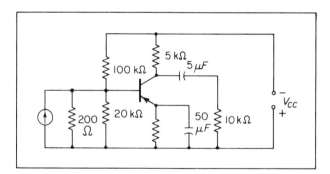

Find: (a) f_{low} (b) f_h (c) $A_{i_{mid}}$ (d) A_i at 250 Hz
(e) A_i at 200 kHz

Solution:

Step 1:   Calculate f_1

$$f_1 = \frac{1}{2\pi R_E C_E}$$

$$f_1 = \frac{1}{2\pi(10)^3 \times 50 \times 10^{-6}}$$

$$f_1 = 3.18 \text{ Hz}$$

Step 2:   Calculate K_0

$$K_0 = 1 + \frac{R_E(1 + h_{fe})}{R_s + h_{ie}}$$

$$K_0 = 1 + \frac{10^3(101)}{200 + 1500}$$

$$K_0 = 1 + 59.4 = 60.4$$

Step 3:   Calculate f_2 or f_{low}

$$f_2 = K_t f_1 = 60.4(3.18)$$

$$f_2 = 192 \text{ Hz}$$

Step 4:   Calculate f_h

$$f_h = \frac{1}{2\pi\left(\dfrac{R_s h_{ie}}{R_s + h_{ie}}\right)C_T}$$

$$f_h = \frac{1}{2\pi(176.5)\ 50 \times 10^{-12}}$$

$$f_h = 18 \text{ MHz}$$

Step 5:   Calculate f_β

$$f_\beta = f_\alpha(1 - |h_{fb}|)$$

$$h_{fb} = \frac{h_{fe}}{1 + h_{fe}}$$

$$f_\beta = 296 \text{ kHz}$$

Step 6:   Calculate $A_{i_{\text{mid}}}$

$$A_{i_{\text{mid}}} = \frac{h_{fe}R_L R_s}{(R_s + h_{ie})(R_L + R_{ie_2})}$$

$$A_{i_{\text{mid}}} = \frac{100(5)(200)}{(200 + 1500)(5 + 10)}$$

$$A_{i_{\text{mid}}} = 3.93$$

Step 7:   Calculate A_i at 250 Hz

$$f_c = \frac{1}{2\pi(R_L + R_{ie_2})C_c}$$

$$f_c = 2.12 \text{ Hz}$$

Since $\dfrac{f}{f_c} \gg 1$ and $\dfrac{f}{f_1} \gg 1$, the resultant equation is equal to:

$$A_{i_{\text{low}}} = \frac{A_{i_{\text{mid}}}\left(j\dfrac{f}{f_2}\right)}{1 + j\dfrac{f}{f_2}}$$

$$A_{i_{\text{low}}} = \frac{3.93(j1.27)}{1 + j\,1.27}$$

$$A_{i_{\text{low}}} = 3.1\ \angle\,38.2°$$

Step 8:   Calculate A_i at 200 kHz

$$A_{i_{\text{hi}}} = \frac{A_{i_{\text{mid}}}}{1 + j\dfrac{f}{f_\beta}}$$

since $\dfrac{f}{f_h} \ll 1.$ $f_h = 18$ MHz

$$A_{i_{\text{hi}}} = \frac{3.93}{1 + j\,.667}$$

$$A_{i_{\text{hi}}} = 3.26\ \angle\,-33.7°$$

A practical two-stage phonograph amplifier is shown in Fig. 4-15. The typical "h parameters" for the transistor are:

$$h_{fe} = 44 \qquad f_\alpha = 3.6\ \text{MHz}$$
$$h_{ie} = 1300\ \Omega \qquad C_T = 35\ \text{pF}$$

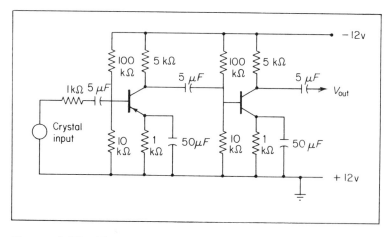

Figure 4-15   *Two-stage Phono Amplifier*

It is left to the student to evaluate the system performance as detailed within the text.

TROUBLESHOOTING ANALYSIS

The various troubles that can occur are: (1) No output, (2) Reduced output or (3) Distorted output. Let us consider the situation of no output. The possible reasons for no output are:

    a. Loss of supply voltage

    b. Open or shorted transistor

    c. Open load or biasing components

    d. Faulty soldering connections

The circuit voltages should be checked and this procedure will normally indicate the exact trouble. If dc voltages are normal, the circuit should be checked for faulty soldering connections.

A reduced or distorted output is usually caused by an aging transistor. Another possibility is the change in the value of the biasing components, which will change the point of operation. In this case, the source power should be removed and a resistance check made using the ohmmeter portion of the voltmeter. An open emitter bypass capacitor can also produce a reduced output.

COMMON SOURCE AMPLIFIER

The basic circuit of a MOSFET RC coupled amplifier is shown in Fig. 4-16.

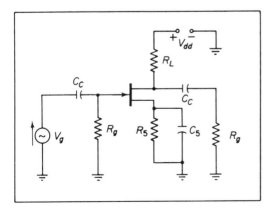

Figure 4-16 *MOSFET RC Coupled Amplifier*

theory   Assume the proper operating potentials have been applied and that the circuit is in a state of equilibrium waiting for signal input.

Step 1:   The signal, V_g, is applied directly to the gate. The sinusoidally varying signal directly varies the drain current in a sinusoidal manner.

Step 2:   The drain to source voltage varies inversely with the applied signal voltage, V_g.

Step 3:   The coupling capacitor, C_c, has two functions; first it blocks dc or prevents the dc voltage from the drain of FET stage one to appear on the gate of stage two. Second, it acts as part of a voltage divider network to the applied sinusoidal voltage. Practical values of coupling capacitors used in RC coupled amplifiers vary between .001 μF and .1 μF.

Step 4:   The gate bias resistor, R_g, closes the dc path from gate to ground.

The analysis of the typical RC coupled MOSFET amplifier proceeds in a manner similar to the bipolar junction transistor. The equivalent circuits for the three audio frequency regions are shown in Fig. 4-17.

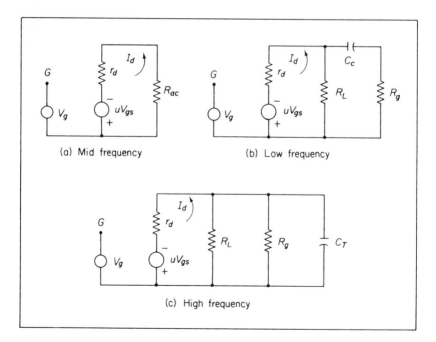

Figure 4-17   *Basic Equivalent Circuits for a MOSFET Amplifier:*
(a) Midfrequency (b) Low Frequency (c) High Frequency

MIDFREQUENCY ANALYSIS

The basic equations for the analysis of the midfrequency region are:

$$uV_g = I_d(r_d + R_{ac})$$

where

$$R_{ac} = \frac{R_L R_g}{R_L + R_g}$$

$$V_0 = -I_d R_{ac}$$

Substituting and simplifying yields the resultant equation.

$$V_0 = -G_m R_e V_g$$

where

$$R_e = \frac{r_d R_L R_g}{r_d R_L + r_d R_g + R_g R_L}; \quad G_m = \frac{u}{r_d}$$

$$A_{v_{mid}} = -G_m R_e$$

LOW FREQUENCY ANALYSIS

Use the Norton's equivalent circuit shown in Fig. 4-18.

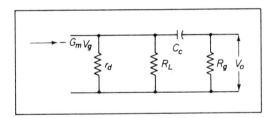

Figure 4-18   *Low Frequency Equivalent Circuit*

$$V_0 = \frac{-G_m V_g \left(\dfrac{R_L r_d}{R_L + r_d}\right) R_g}{\dfrac{R_L r_d}{R_L + r_d} + R_g + \dfrac{1}{j\omega C_c}}$$

Let $R'_e = \dfrac{R_L r_d}{R_L + r_d} + R_g$ and multiply R'_e by R_e. The resultant equation is:

$$R'_e R_e = \frac{R_g R_L r_d}{R_L + r_d}$$

$$A_v = \frac{V_c}{V_g} = \frac{-G_m R'_e R_e}{R'_e + \dfrac{1}{j\omega C_c}}$$

A further simplification results in:

$$\frac{A_{v_{low}}}{A_{v_{mid}}} = \frac{1}{1 + \dfrac{1}{j\omega C_c R'_e}}$$

The value of the half power point frequency f_1 can be determined by:

$$f_1 = \frac{1}{2\pi R'_e C_c}$$

The ratio of the low frequency voltage amplification response with respect to the midfrequency response is then given by:

$$\frac{A_{v_{low}}}{A_{v_{mid}}} = \frac{1}{1 - j\dfrac{f_1}{f}}$$

HIGH FREQUENCY ANALYSIS

Use the Norton's equivalent circuit shown in Fig. 4-19.

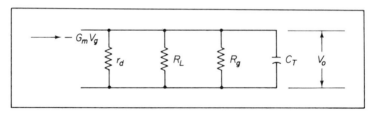

Figure 4-19 *High Frequency Equivalent Circuit*

The three resistors in parallel constitute the single resistor R_e. Thus, the equation for the output voltage is given by:

$$V_0 = -G_m V_g \left(\frac{R_e}{1 + j\omega C_T R_e} \right)$$

and

$$A_{v_{hi}} = \frac{A_{v_{mid}}}{1 + j\omega C_T R_e}$$

The value of the half power point frequency is determined by the formula:

$$f_2 = \frac{1}{2\pi R_e C_T}$$

and

$$\frac{A_{v_{\text{hi}}}}{A_{v_{\text{mid}}}} = \frac{1}{1 + j\dfrac{f}{f_2}}$$

An illustrative problem will demonstrate the theory.

sample problem

The following circuit is given. Determine: (a) f_1, (b) f_2, (c) $A_{v_{\text{mid}}}$, (d) A_v at 40 Hz, (e) A_v at 30 kHz

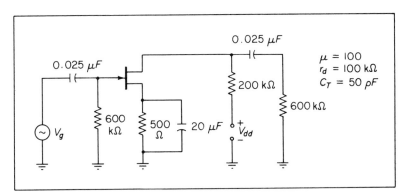

Solution:

    Step 1:   Calculate R_e

$$R_e = \frac{r_d R_g R_L}{r_d R_g + R_g R_L + r_d R_L}$$

$$R_e = 60 \text{ k}\Omega$$

    Step 2:   Calculate R_e'

$$R_e' = R_g + \frac{R_L r_d}{R_L + r_d}$$

$$R_e' = 667 \text{ k}\Omega$$

    Step 3:   Calculate f_1

$$f_1 = \frac{1}{2\pi R_e C_c} = \frac{1}{2\pi (667) 10^3 \times 25 \times 10^{-9}}$$

$$f_1 = 9.54 \text{ Hz}$$

    Step 4:   Calculate f_2

$$f_2 = \frac{1}{2\pi R_e C_T} = \frac{1}{2\pi (60) 10^3 \times 50 \times 10^{-12}}$$

$$f_2 = 53 \text{ kHz}$$

Step 5:   Calculate $A_{v_{mid}}$

$$A_{v_{mid}} = -G_m R_e = -\frac{100}{100} 60$$

$$A_{v_{mid}} = -60$$

Step 6:   Calculate A_v at 40 Hz

$$A_v = \frac{A_{v_{mid}}}{1 - j\frac{f_1}{f}} = \frac{-60}{1 - j\frac{9.54}{40}}$$

$$A_v = \frac{-60}{1 - j.2385}$$

$$A_v = -58.3 \angle 13.4°$$

Step 7:   Calculate A_v at 30 kHz

$$A_v = \frac{A_{v_{mid}}}{1 + j\frac{f}{f_2}} = \frac{-60}{1 + j\frac{30}{53}}$$

$$A_v = -52.2 \angle -29.5°$$

The complete frequency characteristic for a circuit can readily be determined by graphing both the low- and high-frequency characteristic on a single set of coordinates. In addition, the complete phase characteristic for the amplifier can be constructed in a similar manner by combining the low and high frequency characteristic on a single set of characteristics.

Universal curves of relative response and phase shift versus frequency (shown in Fig. 4-20) can be constructed based on the equations derived.

The outstanding features can readily be seen upon investigation of the universal response curve. At frequencies below the lower half power point frequency, f_1, the relative response curve increases at the rate of 6 db per octave or 20 db per decade. The midfrequency region for any amplifier is specified by the rule of thumb relationship. That is:

$$10 f_1 \longleftrightarrow .1 f_2$$

In the high frequency region, the relative response curve decreases at the rate of -6 db/octave or -20 db/decade.

A summary of the number of steps necessary to determine RC coupled amplifier performance is given.

Step 1:   Determine all necessary circuit elements and parameters
Step 2:   Calculate R_e

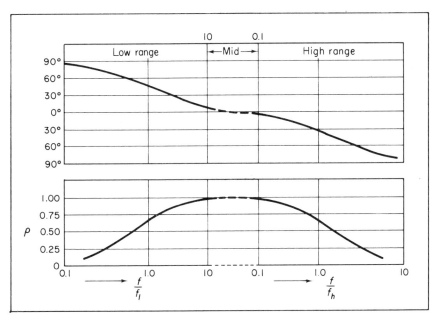

Figure 4-20   *Normalized Gain and Phase Shift Curves Versus Frequency*

Step 3:   Calculate R'_e
Step 4:   Evaluate $A_{v_{mid}}$
Step 5:   Evaluate f_1
Step 6:   Evaluate f_2

A typical MOSFET audio amplifier circuit is shown in Fig. 4-21.

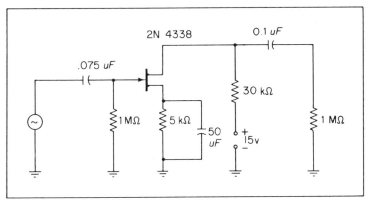

Figure 4-21   *A MOSFET Audio Amplifier Circuit*

FIGURE OF MERIT

A figure of merit for any system is defined by the gain bandwidth product and is very useful in comparing the performance capabilities of any circuit. The voltage amplification value used is the maximum value occurring at midfrequency. Thus, for a bipolar junction transistor, the value of $A_{v_{mid}}$ is given by (for a common emitter circuit):

$$A_{v_{mid}} = \frac{-h_{fe}R_L}{R_g + h_{ie}}$$

If h_{ie} is assumed much larger than R_g then $A_{v_{mid}} = -G_m R_L$ where $G_m = \frac{h_{fe}}{h_{ie}}$. The bandwidth for any amplifier is defined as the difference in the half power point frequencies. The high frequency half power point frequency in an audio amplifier can be considered equal to the total bandwidth. Then,

$$f_h = \frac{1}{2\pi R_L C_T} = f_2$$

The figure of merit symbolized by f_T for the PNP or NPN type transistors is therefore equal to:

$$f_T = A_{v_{mid}} f_h = \frac{G_m}{2\pi C_T}$$

For the FET circuits, the equation for the figure of merit is identical, but the symbol designating the figure of merit is $\mathfrak{M}$. Thus for the FET, the figure of merit is:

$$\mathfrak{M} = \frac{G_m}{2\pi C_T}$$

CASCADED AMPLIFIERS

When a small signal input must be greatly amplified, several stages may be cascaded to provide this greater amplification. The midfrequency amplification of several stages shown in Fig. 4-22 is given by the following relationship for a bipolar junction transistor.

$$A_{i_T} = A_{i_1} \times A_{i_2} \times A_{i_3} \times \cdots \times A_{i_n}$$

If all stages are identical, then

$$A_{i_T} = (A_i)^n$$

where n is the total number of stages.

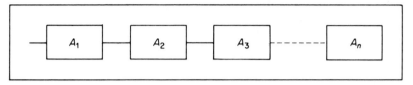

Figure 4-22 *Cascaded Stages Using Block Diagrams*

The same analysis is valid for the FET amplifier system. However, the FET is a voltage activated device; therefore, the system voltage amplification is:

$$A_{vT} = A_{v_1} \times A_{v_2} \times A_{v_3} \times \cdots A_{v_n}$$

If all stages are identical, then

$$A_{v_T} = (A_v)^n$$

where n is the total number of stages.

Cascading of stages results in an increase in system amplification with a consequent reduction in bandwidth. The upper half power point frequency for "n stages," called f_{h_n}, is mathematically related to the half lower point frequency, f_h, for one stage by the equation

$$f_{h_n} = f_h \sqrt{2^{1/n} - 1}$$

The bandwidth reduction factor is defined by:

$$x_{bw_n} = \sqrt{2^{1/n} - 1}$$

Consequently, the value for f_{h_n} is given by:

$$f_{h_n} = x_{bw_n} f_n$$

Similarly, the bandwidth reduction factor for the low frequency region is defined by the relationship:

$$f_{l_n} = \frac{f_l}{x_{bw_n}}$$

A table of values for x_{bw} as a function of the number of stages is given in Table 4-1.

TABLE 4-1 x_{bw} *vs. n* (*BANDWIDTH REDUCTION FACTOR*)

n	x_{bw}
1	1.000
2	.643
3	.51
4	.435
5	.387
6	.350

WIDE BAND AMPLIFIERS

Wide band amplifiers are used to provide uniform amplification from very low frequencies (sometimes from dc) up to the range of 4 MHz. Applications of these circuits are in television receivers, oscilloscopes, radar systems and so forth.

An amplifier configuration capable of wide band amplification is shown in Fig. 4-23. The bipolar junction transistor amplifier or FET amplifier is fed from a voltage source having a low source impedance compared to the transistor input impedance. The load resistor, R_L, is made small so that the transistor frequency response is flat as long as the transistor total conductance remains constant.

One method of extending the high frequency range without reduction

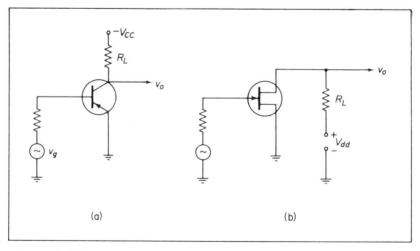

(a)   (b)

Figure 4-23 *Elementary Wide Band Amplifier:* (a) *BJT Circuit* (b) *FET Circuit*

in voltage amplification is to insert a coil in series with R_L as shown in Fig. 4-24. This method is known as shunt peaking.

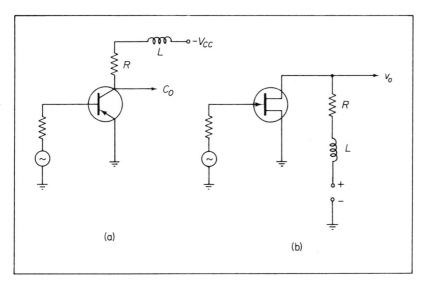

Figure 4-24   *Shunt Peaked Video Amplifier*

The high frequency equivalent circuit for each system is quite similar and is shown in Fig. 4-25.

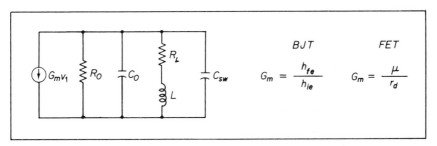

Figure 4-25   *Equivalent Circuit of Shunt Peaked Video Amplifier*

BJT   FET

$$G_m = \frac{h_{fe}}{h_{ie}} \qquad G_m = \frac{u}{r_d}$$

Note that R_o can be defined as the output resistance of the transistor or the dynamic resistance of the FET. It is also evident that C_o is the output capacitance of either device. C_{sw} denotes the stray wiring capacitance. The equivalent circuit can be further simplified as shown in Fig. 4-26.

$$R = R_L + \frac{X_L^2}{R_o}$$

$$C = C_o + C_{sw}$$

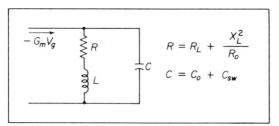

Figure 4-26 *Simplified Equivalent Circuit*

analysis   Using the simplified equivalent circuit, the relative response designated by ρ is equal to:

$$\rho = \frac{A_{v_{hi}}}{A_{v_{mid}}}$$

Since $A_{v_{hi}} = -G_m Z$ and $A_{v_{mid}} = -G_m R$, then

$$\rho = \frac{Z}{R} = \frac{1 + j\frac{\omega L}{R}}{1 - \omega^2 LC + j\omega RC}$$

A family of curves relating the coil inductance to the frequency is shown in Fig. 4-27. The values of k and φ are given by the relationships

$$k = \frac{L}{R^2 C} \quad \text{and} \quad \varphi = \omega RC$$

The phase angle θ for the shunt peaked amplifier is:

$$\theta = \tan^{-1} - [(1 - k)\varphi + k^2 \varphi^3]$$

The graphical solution of this equation is shown in Fig. 4-28. The value of θ is actually the phase angle divided by 360.

It is evident from the graphical results that maximal flatness of response occurs for a value of $k = .414$. In practice, it is simpler to use the value of $k = .5$. An illustrative problem will demonstrate the theory.

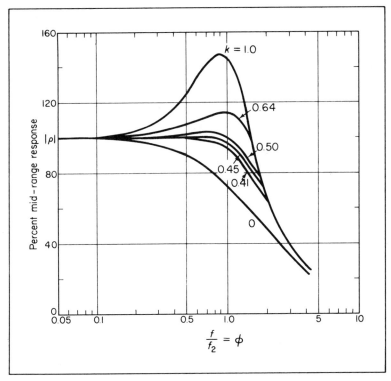

Figure 4-27 *Normalized Response Curves for Shunt Peaked Amplifier*

sample problem

$R = 1 \text{ k}\Omega, C = 40 \text{ pF}, k = .5, \varphi = 1$. Find L and θ.

Solution:

$$L = k R^2 C$$

$$L = .5(10)^6 \times 40 \times 10^{-12}$$

$$L = 20 \ \mu\text{H}$$

From graph for $k = .5$ and $\varphi = 1$, $\theta = .11$

$$\theta = .11 \times 360° = 39.6°$$

A typical FET circuit is shown in Fig. 4-29. This circuit has a voltage amplification of approximately five over a frequency region from 50 Hz to 4 MHz. Short leads must be used to minimize stray wiring capacitances.

An improved method of extending the bandwidth without reduction in amplification is the series shunt compensation method shown in Fig. 4-30.

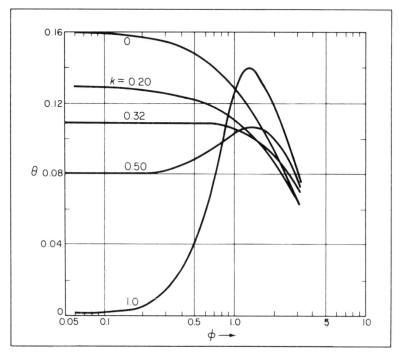

Figure 4-28 *Normalized Time Delay Curves for Shunt Peaking Amplifier*

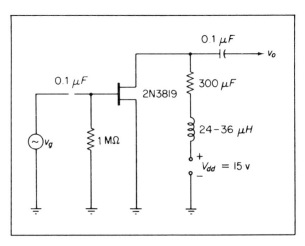

Figure 4-29 *FET Video Amplifier*

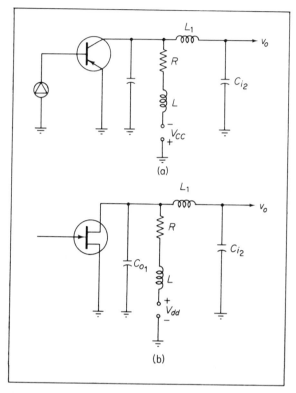

Figure 4-30   *Series Shunt Compensation Video Amplifier: (a) BJT Circuit (b) FET Circuit*

For many semiconductors, flat response over the video region can be obtained with a relatively high load impedance (about 5 kΩ to 10 kΩ) and with loading capacitances (C_{i_2}) on the order of 10 pF. A practical video amplifier is shown in Fig. 4-31.

The amplifier has a high V_{cc} (about 300 V), which allows for the insertion of the 22 kΩ resistor in the emitter circuit. This large resistor selects the operating collector current and provides excellent stability. The 50 μF capacitance across the 22 kΩ resistor presents an ac short circuit so that the emitter resistor does not affect the amplifier performance. The voltage amplification can be maximized when the 1.5 kΩ potentiometer is short circuited. The video signal output voltage is fed directly to the kinescope. The voltage across the 700 Ω resistor is used to provide a video synchronizing signal being fed to the synch separator circuits.

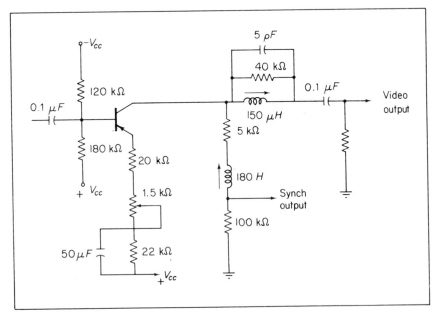

Figure 4-31   *Practical Video Amplifier*

TUNED VOLTAGE AMPLIFIER

Both radio and TV receivers always have tuned circuit amplifiers that serve to amplify a band of frequencies and reject undesirable frequencies. The advantages of these circuits are:

1. Higher voltage amplification with a corresponding narrower bandwidth

2. Better sensitivity

3. Lower signal to noise ratio at the input to the receiver

The properties of these amplifiers depend on the characteristics of resonant circuits. A review of the necessary portions of series and parallel resonant theory will be presented briefly. Consider the series RLC circuit shown in Fig. 4-32.

analysis For sinusoidal waveforms, the current in the circuit at any frequency is:

$$I = \frac{V}{R + j(X_L - X_c)}$$

where $X_L = \omega L$

$\qquad X_c = \dfrac{1}{\omega C}$

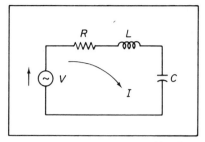

Figure 4-32 *Series RLC Circuit*

The circuit is considered resonant when the j term is zero. Setting the j term equal to zero results in:

$$\omega L = \frac{1}{\omega C}$$

$$f_r = \frac{1}{2\pi\sqrt{LC}}$$

At resonance, the impedance in the circuit is given by R. The current in the circuit at resonance is:

$$I_r = \frac{V}{R}$$

The circuit response is defined as the ratio of the current existent at any frequency with respect to that at resonance. Thus,

$$\frac{I}{I_r} = \frac{R}{R + j(X_L - X_c)}$$

A graph of the relative response versus frequency curve is shown in Fig. 4-33. The frequencies f_1 and f_2 are known as the half power point frequencies and are used to define the bandwidth of the circuit.

The Q of the circuit is defined by the ratio of inductive reactance to resistance. Thus,

$$Q = \frac{\omega L}{R}$$

The circuit Q is a measure of the selectivity of the circuit. The higher the Q, the sharper the bandwidth. The sharpness of resonance for a series circuit is given by the mathematical relationship:

$$\frac{f_2 - f_1}{f_r} = \frac{1}{Q}$$

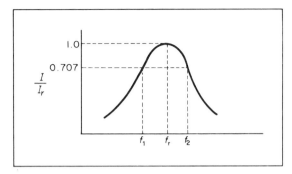

Figure 4-33   *Relative Current Response of a Series RLC Circuit*

The reactance versus frequency characteristic for a series circuit is shown in Fig. 4-34.

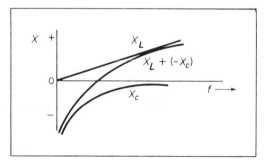

Figure 4-34   *Reactance vs Frequency Curves for a Series RLC Circuit*

It is evident that at frequencies below resonance the circuit is capacitive in nature, whereas for frequencies above resonance, the circuit is inductive in nature. Consequently, in many circuit applications, the purpose of the tuned circuit is to tune in the station or carrier by using either a variable L or C for the job.

The circuit response at all frequencies can be determined as a function of circuit current at resonance by the mathematical relationship:

$$\rho = \frac{I}{I_r} = \frac{1}{1 + j\left(\dfrac{1}{\omega RC}\right)(\omega^2 LC - 1)}$$

If the frequency considered is off resonance by a fractional amount, specified by delta (δ), then let the frequency equal:

$$f = f_r(1 + \delta)$$

Substituting and simplifying the expression for circuit response yields

$$\rho = \frac{1}{1 + jQ\left(\frac{2\delta + \delta^2}{1 + \delta}\right)}$$

If δ is assumed equal to or less than .1, then

$$\rho = \frac{1}{1 + j2\delta Q}$$

PARALLEL RESONANCE

A parallel combination of RL and C is shown in Fig. 4-35.

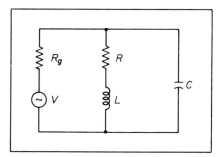

Figure 4-35 *Parallel RLC Network*

The total input admittance (Y) is equal to:

$$Y = \frac{1}{R + j\omega L} + j\omega C$$

or

$$Y = \frac{R}{R^2 + \omega^2 L^2} - j\frac{\omega L}{R^2 + \omega^2 L^2} + j\omega C$$

Resonance occurs when the total susceptance equals zero. Thus,

$$B_L = B_c$$

$$\frac{\omega L}{R^2 + \omega^2 L^2} = \omega C$$

Solving this equation for the frequency at resonance yields

$$f_{ar} = \frac{1}{2\pi\sqrt{LC}}\sqrt{1 - \frac{R^2 C}{L}}$$

The conductance of the circuit at resonance is:

$$G = \frac{R}{R^2 + \omega^2 L^2}$$

The reciprocal of conductance is resistance. The resistance of the circuit at resonance is given by:

$$R_{ar} = R(1 + Q^2)$$

If the Q of the coil is equal to or greater than ten, the following relationships are valid:

$$R_{ar} = Q^2 R = QX_L = QX_c = \frac{L}{CR}$$

The sharpness of resonance for a parallel resonant circuit is given by the mathematical relationship:

$$\frac{f_2 - f_1}{f_{ar}} = \frac{\Delta f}{f_{ar}} = \frac{1}{Q_L} + \frac{X_L}{R_g}$$

In the special case in which the generator resistance is equal to the resistance of the parallel RLC circuit at resonance, then the sharpness of resonance becomes

$$\frac{f_2 - f_1}{f_{ar}} = \frac{2}{Q_L}$$

The impedance of the circuit at all frequencies is:

$$Z = \frac{(R + j\omega L)\left(\dfrac{1}{j\omega C}\right)}{R + j\omega L + \dfrac{1}{j\omega C}}$$

$$Z = \frac{\dfrac{L}{C} - j\dfrac{R}{\omega C}}{R + j\omega L + \dfrac{1}{j\omega C}}$$

Factoring L/C from the numerator and R from the denominator yields the resultant equation

$$Z = \frac{L}{CR}\left[\frac{1 - j\dfrac{R}{\omega L}}{1 + j\left(\dfrac{\omega^2 LC - 1}{\omega RC}\right)}\right]$$

A graph of impedance versus frequency is shown in Fig. 4-36.

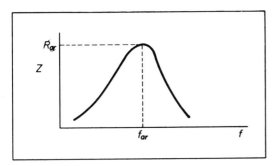

Figure 4-36   *Impedance Versus Frequency*

At frequencies near resonance, let $f = f_{ar}(1 + \delta)$; then the equation for the impedance of the circuit becomes

$$Z = \frac{L}{RC}\left[\frac{1 - j\dfrac{1}{Q(1 + \delta)}}{1 + jQ\left(\dfrac{2\delta + \delta}{1 + \delta}\right)}\right]$$

If the following assumptions are valid:

$$\delta \leq .1 \quad \text{and} \quad Q \geq 10$$

the equation simplifies to:

$$Z = R_{ar}\left(\frac{1}{1 + j2\delta Q}\right)$$

A parallel RLC combination of the type given can be replaced by three elements as shown in Fig. 4-37.

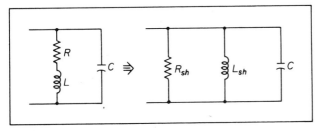

Figure 4-37   *Equivalent Networks*

The values of R_{sh} and L_{sh} can be determined from the following analysis.

$$\frac{1}{R^2 + j\omega L} = \underbrace{\frac{R}{R^2 + X_L^2}}_{G_{sh}} - j\underbrace{\frac{X_L}{R^2 + X_L^2}}_{B_{sh}}$$

$$R_{sh} = \frac{1}{G_{sh}} = \frac{R^2 + X_L^2}{R}$$

$$X_{L_{sh}} = \frac{1}{B_{sh}} = \frac{R^2 + X_L^2}{X_L}$$

If the Q of the coil is equal to or greater than ten, then

$$R_{sh} = QX_L = R_{ar}$$
$$L_{sh} = L$$

This resultant equation indicates that the series coil resistance can be removed and replaced by a resistor in parallel and equal to the tank or tuned circuit impedance at resonance.

SINGLE TUNED DIRECT COUPLED AMPLIFIERS

In many electronic systems, the radio frequency (RF) signal input is generally extremely small. Before intelligence can be removed from the weak signal, amplification of signal level is required. This function is performed by an RF amplifier.

Note that since the signal level is small (on the order of a few microvolts), the RF amplifier is called a small signal amplifier. In addition, it is evident that the power level is negligible and that the main requirement of the amplifier is to increase the amplitude of the signal. Consequently, this type of amplifier is also called an RF or tuned voltage amplifier.

The major points of interest in RF amplifiers are: first, what is the maximum voltage amplification of the stage? The second point is, what band of frequencies will the amplifier permit to pass through? The amplification of the amplifier at any frequency within the pass band region of the amplifier is the third point of interest.

It is evident that the signal output must be coupled from one stage to the next. The system requirements for the coulping circuit are:

1. To allow the radio frequencies or the signal free passage without losses.

2. To prevent any dc component from one stage to pass through to the next stage.

The basic circuit of a single tuned direct coupled amplifier (STDC) is shown in Fig. 4-38.

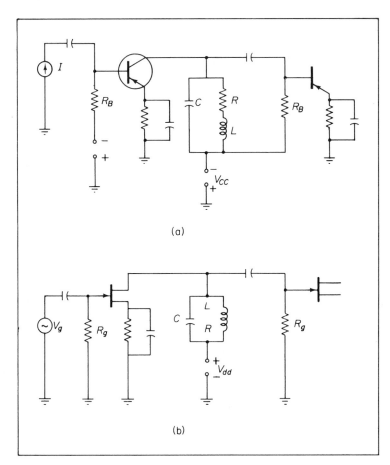

Figure 4-38 *STDC (a) BJT (b) FET Versions*

The equivalent circuit of the bipolar junction transistor that is used for radio frequency circuit analysis is shown in Fig. 4-39.

It is evident that a feedback path exists from the collector to the base and may be adequate to sustain oscillation within the circuit. Consequently, the method used to eliminate or neutralize the feedback path is called "unilateralization." The inductor in the tuned circuit is tapped to form a bridge network as shown in Fig. 4-40.

The resultant circuit illustrating the unilateralization network is shown in Fig. 4-41.

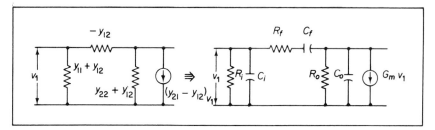

Figure 4-39   *Equivalent Circuit of a BJT Transistor*

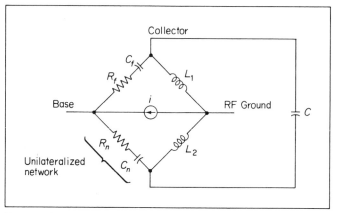

Figure 4-40   *Unilateralization of BJT Amplifier*

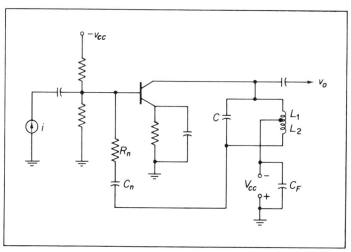

Figure 4-41   *Unilateralized BJT STDC Amplifier*

With the application of the unilateralization network, the equivalent circuits for both the BJT and the FET amplifiers become almost identical. The essential difference is the method of determining G_m. The equivalent circuit of the amplifier is shown in Fig. 4-42.

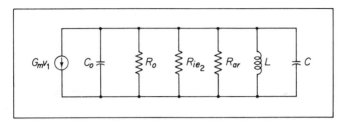

Figure 4-42 *Simplified Equivalent Circuit for STDC Amplifier*

BJT
$$G_m = |(y_{21} - y_{12})|$$
FET
$$G_m = \frac{u}{r_d}$$
$$R_o = r_d$$

analysis   At resonance the circuit is purely resistive. Thus,

$$V_{o_r} = -G_m v_1 R_s$$

where

$$R_e = \frac{1}{\dfrac{1}{R_o} + \dfrac{1}{R_{ie_2}} + \dfrac{1}{R_{ar}}}$$

The voltage output at any frequency is given by:

$$V_o = -G_m v_1 Z$$

where

$$Z = \frac{R_e(jX_L)(-jX_c)}{jX_L R_e - jR_e X_c + (jX_L)(-jX_c)}$$

The voltage amplification of the circuit at resonance is defined by the ratio of V_{o_r} to v_1, whereas the voltage amplification of the circuit at any frequency is given by the ratio of V_o to v_1. The relative response of any circuit is defined by the ratio of the voltage amplification at any frequency with respect to that at resonance. Thus,

$$\rho = \frac{A_v}{A_{v_{res}}} = \frac{1}{1 + j\dfrac{X_L - X_c}{R_e}}$$

The effective Q (Q_e) of the circuit is defined in terms of the unloaded Q. Thus,

$$Q_e = \frac{R_e}{X_L} = Q\frac{R_e}{R_{ar}}$$

By letting $x = 2\sigma Q_e$, the resultant value of ρ for the circuit becomes

$$\rho = \frac{1}{1 + jx}$$

It is evident that the relative response is equal to 0.707 when x is equal to unity. This value of x is called the bandwidth value and is symbolized by x_{bw}. Thus,

$$x_{bw} = 1, \qquad 1 = Q_e\frac{\Delta f}{f_{ar}}$$

and

$$Q_e = \frac{f_{ar}}{\Delta f}$$

A graph of the relative response as a function of x is shown in Fig. 4-43.

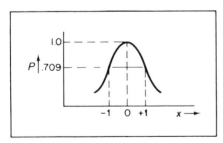

Figure 4-43 *Relative Response for a STDC Amplifier*

The MOSFET transistor is far superior to the BJT when used as an RF amplifier. This transistor has a very low feedback or reverse transfer capacitance. Consequently, no special neutralizing circuitry is required. The dc operating point is established by proper values of R_s and bypass capacitor C_s.

An illustrative problem will demonstrate the theory.

sample problem

An FET STDC has the following given data:

$$r_d = 1 \text{ M}\Omega \qquad Q = 79.5 \qquad f_{ar} = 1 \text{ MHz}$$
$$G_m = 2000 \ \mu\text{mhos} \qquad L = 500 \ \mu\text{H} \qquad R_g = 300 \text{ k}\Omega$$

Find: (a) $A_{v_{res}}$ (b) Δf (c) the frequencies where $\rho = 0.75$

Solution:

Step 1:

$$R_{ar} = QX_L$$
$$R_{ar} = 79.5 \times 2\pi \times 10^6 \times 500 \times 10^{-6}$$
$$R_{ar} = 250 \text{ k}\Omega$$

Step 2:

$$R_e = \frac{R_{ar}R_g r_d}{R_{ar}R_g + R_{ar}r_d + R_g r_d}$$
$$R_e = 120 \text{ k}\Omega$$

Step 3:

$$A_{v_{res}} = 2 \times 10^{-3} \times 120 \times 10^3 = 240$$

Step 4:

$$Q_e = Q\frac{R_e}{R_{ar}} = 79.5\frac{120}{250}$$
$$Q_e = 38.2$$

Step 5:

$$\Delta f = \frac{f_{ar}}{Q_e} = \frac{10^6}{37.68}$$
$$\Delta f = 26.0 \text{ kHz}$$

Step 6:

$$\rho^2 = \frac{1}{1 + x^2}$$
$$x^2 = \frac{1}{\rho^2} - 1$$
$$x^2 = \frac{1}{.5625} - 1$$
$$x = .88$$
$$df = 22.88 \text{ kHz}$$
$$f_b = 1.0114 \text{ MHz}$$
$$f_a = .9886 \text{ MHz}$$

DOUBLE TUNED TRANSFORMER COUPLED AMPLIFIER

A circuit using a double tuned transformer coupling is shown in Fig. 4-44. Circuits of this type are commonly used in radio or TV receivers for intermediate frequency amplification.

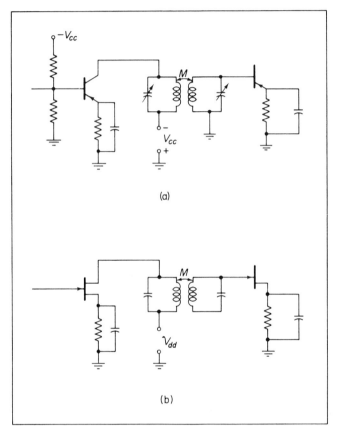

(a)

(b)

Figure 4-44   *Double Tuned Transformer Coupled Amplifier:* (a) *BJT DTTC Amplifier* (b) *FET DTTC Amplifier*

Once the BJT amplifier circuit has been unilateralized, the equivalent circuits become identical as shown in Fig. 4-45:

$$C_{e_1} = C_o + C_1$$

BJT   FET

$$R = R_o \quad R = r_d$$

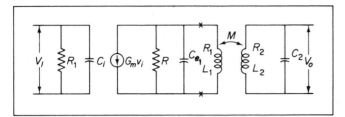

Figure 4-45 *Equivalent Circuit of DTTC Amplifier*

analysis Assume that R is much greater than X_{c_1}. Then, Thevenize the circuit as shown in Fig. 4-46.

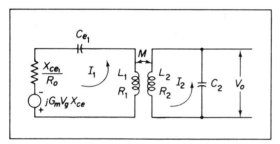

Figure 4-46 *Simplified Equivalent Circuit*

The general two-loop network equations are for this circuit:

$$V_1 = I_1 Z_{11} - I_2 Z_m$$
$$V_2 = -I_1 Z_m + I_2 Z_{22}$$

where:

$$V_1 = jG_m V_g X_{ce_1} \qquad V_2 = 0$$

$$Z_{11} = R_1 + \frac{X_{ce_1}^2}{R} + j(X_{L_1} - X_{ce_1})$$

$$Z_m = jX_m; \qquad Z_{22} = R_2 + j(X_{L_2} - X_{c_2})$$

Solving for the secondary current yields:

$$I_2 = \frac{Z_m V_1}{Z_{11} Z_{22} - Z_m^2}$$

The output voltage is:

$$V_o = -I_2(-jX_{c_2}) = \frac{Z_m jX_{c_2} V_1}{Z_{11} Z_{22} - Z_m^2}$$

The voltage amplification of the circuit is given by the following equation:

$$A_v = \frac{V_0}{V_1} = \frac{-jG_m X_{c_1} X_{c_2} X_m}{\left[R_1 + \frac{X_{c_1}^2}{R} + j(X_{L_1} - X_{c_1}) \right][R_2 + j(X_{L_2} - X_{c_2})] + X_m^2}$$

The following definitions will be used for simplification of circuit analysis:

$$R_{1_T} = R_1 + \frac{X_{c_1}^2}{R} \qquad\qquad \gamma = kQ_g$$

$$Q_{e_1} = \frac{R_{e_1}}{X_{L_1}} = \frac{\omega_{ar} L_1}{R_{1_T}} \qquad\qquad a = \frac{Q_a}{Q_g}$$

$$Q_{e_2} = \frac{R_{e_2}}{X_{L_2}} = \frac{\omega_{ar} L_2}{R_2} \qquad\qquad x = Q_g \frac{df}{f_{ar}}$$

$$Q_g = \sqrt{Q_{e_1} Q_{e_2}} \qquad\qquad M = k\sqrt{L_1 L_2}$$

$$Q_a = \frac{Q_{e_1} + Q_{e_2}}{2}$$

Substituting these definitions into the equation for voltage amplification:

$$A_v = \frac{G_m Q_{e_1} Q_{e_2} \omega_{ar} M}{\left(1 + jQ_{e_1}\frac{df}{f_{ar}}\right)\left(1 + jQ_{e_2}\frac{df}{f_{ar}}\right) + \gamma^2}$$

This equation can be simplified further by algebraic manipulation to the resultant form in terms of magnitude:

$$A_v = \frac{G_m \gamma Q_g \omega_{ar} \sqrt{L_1 L_2}}{\sqrt{x^4 - 2x^2(\gamma^2 + 1 - 2a^2) + (\gamma^2 + 1)^2}}$$

A graph of the relative response versus frequency is shown in Fig. 4-47. Note that the relative response varies as a function of the degree of coupling or γ. The critical value of γ is defined by:

$$\gamma_{crit} = \sqrt{2a^2 - 1}$$

The mathematical expression for the relative response of the circuit for critical coupling or overcoupling is given by:

$$\rho = \frac{2a\sqrt{\gamma^2 + 1 - a^2}}{\sqrt{x^4 - 2x^2(\gamma^2 + 1 - 2a^2) + (\gamma^2 + 1)^2}}$$

It should be noted that a special case exists whereby the loaded Q's of the primary and secondary circuits are equal. This, of course, sets the

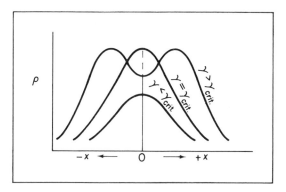

Figure 4-47 *Response Characteristics of a DTTC Amplifier for Different Values of Coupling*

geometric Q equal to the arithmetic Q and critical coupling occurs when γ is equal to unity. The resultant formulas are obtained by setting a equal to unity and solving. Thus, the relative response is:

$$\rho = \frac{2\gamma}{\sqrt{x^4 - 2x^2(\gamma^2 - 1) + (\gamma^2 + 1)^2}}$$

The bandwidth of a different Q circuit is determined by setting the relative response equal to 0.707 and solving for x_{bw}. Thus,

$$x_{bw} = \sqrt{(\gamma^1 + 1 - 2a^2) + 2a\sqrt{\gamma^2 + 1 - a^2}}$$

An illustrative problem will demonstrate the theory.

sample problem

A transistorized double tuned transformer coupled amplifier is center tuned to a frequency of 50 MHz. The transistor parameters are:

$$y_{11} = (25 + j15)10^{-3} \text{ mhos}$$
$$y_{22} = (2 + j6)10^{-3} \text{ mhos}$$
$$y_{12} = -(1 + j0.4)10^{-3} \text{ mhos}$$
$$y_{21} = (40 - j90)10^{-3} \text{ mhos}$$

The primary and secondary Q terms are equal ($a = 1$). The value of γ is 3. The unloaded coil Q in the primary is 50 and the coil inductance is $80 \times 10^{-3}\ \mu H$. Find:

(a) the values of R_n and C_n

(b) the circuit bandwidth

(c) the maximum voltage amplification

(d) the voltage amplification at a frequency of 56.5 MHz.

Solution:

    Step 1:   Convert y_{12} to an equivalent circuit

$$z_f = \frac{10^3}{1 + j\,.4} = 860 - j\,344$$

    If $L_1 = L_2$,   then $R_n = 860\ \Omega$

$$C_n = 9.25\ \text{pF}$$

    Step 2:   Calculate R_o and C_o

$$y_o = y_{22} + y_{12} = (1 + j\,5.6)10^{-3}$$

$$R_o = 1000\ \Omega \quad C_o = 17.8\ \text{pF}$$

    Step 3:   Calculate R_{ar} of the primary

$$R_{ar} = QX_L = 50 \times 2\pi \times 50 \times 10^6 \times 80 \times 10^{-9}$$

$$R_{ar} = 1256\ \Omega$$

    Step 4:   Calculate R_e

$$R_e = \frac{R_{ar}R_o}{R_{ar} + R_o} = 510\ \Omega$$

    Step 5:   Calculate G_m

$$G_m = y_{21} - y_{12} = (41 - j\,89.6)10^{-3}$$

$$G_m = 98.6\ \angle\,{-65.4°}\ 10^{-3}\ \text{mhos}$$

    Step 6:   Calculate $A_{v_{\text{max}}}$

$$A_{v_{\text{max}}} = \frac{G_m R_e}{2} = 25.05$$

    Step 7:   Calculate Q_e

$$Q_e = \frac{R_e}{X_L} = 20.3$$

    Step 8:   Calculate x_{bw}

$$x_{bw} = \sqrt{\gamma^2 + 2\gamma - 1} = 3.74$$

    Step 9:   Calculate Δf

$$\Delta f = x_{bw}\frac{f_0}{Q_e} = 9.2\ \text{MHz}$$

Step 10:   Calculate ρ

$$x = Q_e \frac{df}{f_{ar}} = 3.65$$

$$x^2 = 13.3 \qquad x^4 = 174$$

$$\rho = \frac{2\gamma}{\sqrt{x^4 + 2x^2 \, (\gamma^2 - 1) + (\gamma^2 + 1)^2}}$$

$$\rho = .768$$

Step 11:   Calculate A_v at 54.5 MHz

$$A_v = \rho A_{v_{max}} = 19.2$$

TROUBLESHOOTING ANALYSIS

Various conditions can result in loss of output. These are:

(a) lack of input signal

(b) loss of supply voltage

(c) improper bias

(d) tuned circuit component failure

(e) defective transistor

Note that an open source or emitter resistor results in no output. On the other hand, an open source or emitter by-pass capacitor will produce degeneration resulting in reduced output. Changing value of bias resistors as the result of heat or increased temperature will produce a distorted output.

Check all bias and supply voltages with a voltmeter to establish circuit performance. It must be remembered that the interelement capacitances can have drastic effects on circuit performance and extreme care must be used when replacing transistors. Detuning will cause a *no output* condition if the improper transistor is used. This condition is evident when dc voltages are near normal but, although the signal is applied, there is no output.

TROUBLESHOOTING FET AMPLIFIERS

Amplifier malfunctions usually fall into two categories: those that change the dc voltages and those that do not. Circuit faults that do not change the dc voltages usually affect the signal path. The defective component that causes this condition can usually be found by signal tracing or by logically examining the circuit and systematically eliminating the components that can change the dc voltages. The elements that remain can readily be checked by a multimeter for proper values.

Voltage measurements are generally one of the best methods to use in troubleshooting a circuit. However, the technician must know what the correct voltage should be at each point in the circuit under test. There are two practical methods to determine the proper voltage readings. The first and best method is to obtain a schematic diagram of the equipment under test and compare the measured values against the original test values marked on the schematic. If a schematic diagram for the particular equipment under test is not available any similar diagram can be used.

On most schematic diagrams, the *small print* usually specifies the type of voltmeter that was used to measure the indicated voltages, whether there was an input signal, what control or controls were set in a specific way and other types of information. In addition, it is usually noted that although the input ac voltage is fixed, the measured voltages may vary by plus or minus 20 percent ($\pm 20\%$).

The use of a lower impedance voltmeter rather than the type specified may result in low voltage readings throughout although the amplifier may be operating properly. It should also be noted that if the ac voltage input (line voltage) varies, then the deviation in the measured voltage readings may be greater than the plus or minus 20 percent even though the amplifier is performing properly.

For troubleshooting analysis, refer to Fig. 4-48, which illustrates an FET amplifier.

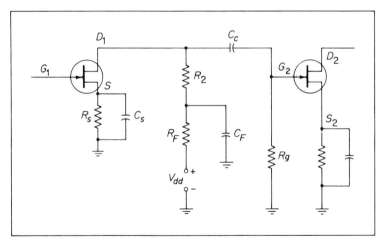

Figure 4-48   *Typical FET Amplifier*

Consider the circuit problem in which the voltage measured from drain to ground is lower than the required value. This condition may be caused by:

1. A decreased voltage supply (V_{dd}).

2. Incorrect voltage divider relationship between R_L and R_F.

3. Incorrect operating bias.

The first step in the troubleshooting procedure is to check the power supply voltage. If a low reading is obtained, then the power supply should be repaired or replaced. However, if the voltage reading is correct, the trouble probably is with the $R_F C_F$ decoupling filter. Note: the voltmeter can show a low reading of voltage at the junction of R_L and R_F, but a correct reading across the power supply. This can be caused by either an increased value of R_F or by a leaky or shorted capacitor C_F. Remove the voltage V_{dd} and check R_F with an ohmmeter. If satisfactory, check C_F for leaks by using the ohmmeter.

Another factor that might cause a low drain voltage is an increase in the value of R_L. Heat and normal aging may cause a change in R_L. The value of R_L can also be checked with an ohmmeter.

A decrease in source resistance (R_s) or a shorted or leaky bypass capacitor (C_s) may also yield a low drain voltage. Generally, because of the relatively low voltages at which these capacitors operate, shorted source bypass capacitors are rare. These components can readily be checked with an ohmmeter. In the case of the capacitor, remember to disconnect one side of the capacitor first.

A defective or failing FET can cause a low drain voltage. If the trouble is due to a poor FET, substitution with an FET known to be good will eliminate this possibility. A shorted or leaky bypass capacitor (C_s) can cause low drain voltage. It will lower the drain voltage of the FET to whose drain it is connected by causing additional current to flow through the resistor, R_L. It will also lower the drain voltage of the FET to whose gate it is connected by driving the gate positive and increasing the gate current of the FET. A method of locating a shorted coupling capacitor is to disconnect the capacitor at the gate side. Connect a dc voltmeter from the gate end of the capacitor to a ground point on the chassis. Apply V_{dd}. If the capacitor is leaky, the needle on the voltmeter will deflect appreciably. Generally, a replacement capacitor will correct the defect.

In the amplifier of Fig. 4-48 the gate and source are returned to ground. Consider the case whereby the gate and source voltages are returned to a negative voltage instead of to ground. Under these conditions, it is possible for the drain to source voltage to be negative. This condition may be caused by a low source bias voltage, which may be caused by a shorted bypass capacitor (C_s) or a leaky coupling capacitor (C_s). Note that either condition can produce excessive drain current. Another reason for the drain voltage to be negative is that the load resistor (R_L) may be open. This condition can readily be checked with an ohmmeter.

Consider the circuit problem in which the voltage measured is higher than normal. The possible causes are:

1. Defective FET.
2. Source bias increased due to increased value of source resistance (R_s).
3. Decreased value of load resistor (R_L).
4. Decreased value of drain decoupling resistor (R_F).

One possible cause of a high drain voltage is an increase of power supply voltage. This condition can rapidly be checked. Consider the case where the drain voltage is equal to the supply voltage. Under these conditions the drain current must be zero, which can occur only if (a) R_L is shorted or (b) FET is open. The value of R_L can be checked with an ohmmeter and if the resistor proves good, replace the FET.

Open bypass and coupling capacitors are examples of component failures that will not affect the dc portions of the circuit. Open capacitors are simple to locate since the full signal is present on one side, but there is no signal on the other side.

problems

1. An RC coupled amplifier has the following given data:

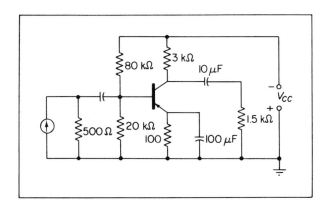

$h_{ie} = 2\ \text{k}\Omega$

$h_{fe} = 100$

$C_T = 50\ \text{pF}$

$f_\alpha = 30\ \text{MHz}$

Find: (a) $A_{i_{\text{mid}}}$ (b) f_l (c) f_h

2. An RC coupled amplifier has the following given data:

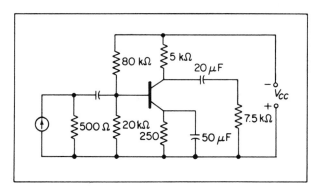

$h_{ie} = 1.5\ \text{k}\Omega$

$h_{fe} = 70$

$C_T = 50\ \text{pF}$

$f_\alpha = 20\ \text{MHz}$

Find: (a) $A_{i_{\text{mid}}}$ (b) f_l (c) f_h

3. Two stages having individual stages equal to that of problem 1 are combined. Determine:
   (a) $A_{i_{midT}}$ (b) f_{l_T} (c) f_{h_T}

4. Two stages having individual stages equal to that of problem 2 are combined. Determine:
   (a) $A_{i_{midT}}$ (b) f_{l_T} (c) f_{h_T}

5. The circuit shown has three separate inputs. Prove that the output signal, $V_o = k(V_1 + V_2 + V_3)$.
   Assume that the u's and r_d's are all identical.

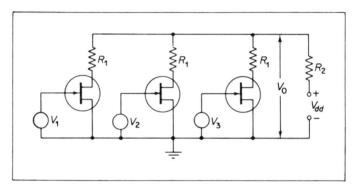

6. Determine the value of the output signal in terms of both input signals. Assume that the FET parameters are all equal.

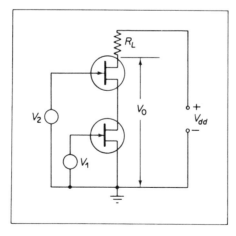

7. Determine the value of the transfer function when (a) $V_1 = 0$ (b) $V_2 = 0$. Assume the FET parameters equal.
   $u = 20$
   $r_d = 10 \text{ k}\Omega$

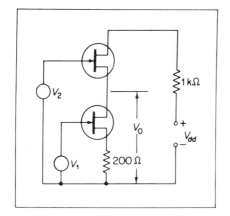

8. Determine the value of V_o for the given circuit.

$u = 100$
$r_d = 80 \text{ K}$
$R_L = 75 \text{ K}$
$R_1 = 2 \text{ K}$
$R_2 = 10 \text{ K}$
$V_1 = 1.5 \text{ V}$
$V_2 = 3 \text{ V}$

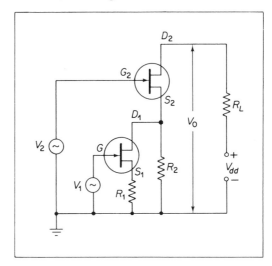

9. Determine the value of V_{o_1} and V_{o_2} in terms of V_{in}.

$u = 20$
$r_d = 40 \text{ K}$
$R_L = 70 \text{ K}$
$R_s = 2 \text{ K}$

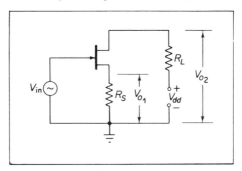

10. An RC coupled FET amplifier has the following given data:
$R_L = 100 \text{ k}\Omega$ $C_c = .005 \ \mu\text{F}$ $G_m = 1600 \ \mu\text{mhos}$
$R_g = 250 \text{ k}\Omega$ $C_T = 40 \text{ pF}$ $u = 100$
Find: (a) f_l (b) f_h (c) $A_{v_{mid}}$ (d) A_v at $f = 200$ Hz
(e) A_v at $f = 100$ kHz.

11. An RC coupled amplifier using an FET has the following given data:
$R_L = 75 \text{ k}\Omega$ $C_T = 75 \text{ pF}$ $u = 75$
$R_g = 150 \text{ k}\Omega$ $r_d = 50 \text{ k}\Omega$
Find: (a) $A_{v_{mid}}$ (b) f_h (c) C_c for $f_l = 25$ Hz.

12. The bandwidth of an FET amplifier is from 20 Hz to 30 kHz. Determine the frequency region over which the voltage ampliﬁciation is equal to .8 of the midfrequency region.

13. An RC coupled FET amplifier has the following given data:
$R_L = 50 \text{ k}\Omega$ $C_c = .05 \ \mu\text{F}$ $u = 100$
$R_g = 200 \text{ k}\Omega$ $C_T = 150 \text{ pF}$ $r_d = 100 \text{ k}\Omega$
Find: (a) $A_{v_{mid}}$ (b) f_l (c) f_h (d) The voltage amplification where the phase shift through the output circuits is $\pm 35°$.

14. An RC coupled FET amplifier has the following given data:
$R_L = 100 \text{ k}\Omega$ $C_c = .025 \ \mu\text{F}$ $u = 100$
$R_g = 300 \text{ k}\Omega$ $C_T = 40 \text{ pF}$ $r_d = 75 \text{ k}\Omega$
Find: (a) $A_{v_{mid}}$ (b) f_l (c) f_h (d) A_v at $f = 20$ Hz
(e) A_v at $f = 100$ kHz.

15. A single tuned direct coupled amplifier has the following given data:
$y_{11} = (10 + j\,3)10^{-3} \text{ mhos}$ $f_{ar} = 100$ MHz
$y_{12} = (-(.5 + j\,.25)10^{-3} \text{ mhos}$ $Q = 100$
$y_{21} = (.5 + j\,.75)10^{-3} \text{ mhos}$ $L_1 = 1 \ \mu\text{H}$
$y_{22} = (.505 + j\,1.25)10^{-3} \text{ mhos}$
Find: (a) R_n and C_n
(b) $A_{v_{max}}$
(c) the circuit bandwidth
(d) the relative response at 101 MHz.

16. A single tuned direct coupled amplifier has the following given data:
$y_{11} = (4 + j\,7)10^{-3} \text{ mhos}$ $f_{ar} = 25$ MHz
$y_{12} = -j\,2 \times 10^{-3} \text{ mhos}$ $\Delta f = 0.5$ MHz
$y_{21} = (6 - j\,10)10^{-3} \text{ mhos}$ $C_t = 100$ pF
$y_{22} = (.1 + j\,3)10^{-3} \text{ mhos}$
Find: (a) R_n and C_n
(b) the maximum voltage amplification
(c) the relative response at 25.15 MHz

17. A single tuned direct coupled amplifier has the following given data:

$G_m = 2000$ μmhos $\qquad R_g = 100$ kΩ $\qquad Q_o = 200$

$r_d = 500$ kΩ $\qquad C_t = 20$ pF $\qquad \Delta f = 200$ kHz

Find: (a) L

(b) $A_{v_{max}}$

(c) the value of relative response at $f_{ar} + 40$ kHz

18. A single tuned direct coupled amplifier has the following given data:

$G_m = 2000$ μmhos $\qquad Q = 100$ $\qquad R_g = 300$ kΩ

$r_d = 150$ kΩ $\qquad R_{coil} = 5$ Ω $\qquad C_t = 159$ pF

Find: (a) f_{ar}

(b) the circuit bandwidth

(c) the voltage amplification at a frequency of $f_{ar} + 15$ kHz.

19. A double tuned transformer-coupled amplifier has the parameters specified in problem 15. The center frequency is 100 MHz. Other given data:

$L_1 = L_2 = 1$ μH $\qquad Q_1 = 100$ $\qquad a = 1$ $\qquad \gamma = 3$

Find: (a) the maximum voltage amplification

(b) the circuit bandwidth

(c) the relative response of the circuit at 101 MHz.

20. A double tuned transformer-coupled amplifier has the parameters specified in problem 16. The center frequency is 25 MHz. Other given data:

$\Delta f = 0.5$ MHz $\qquad C_t = 100$ pF $\qquad a = 1$ $\qquad \gamma = 2$

Find: (a) the maximum voltage amplification

(b) the relative response of the network at 25.2 MHz

21. A double tuned transformer-coupled amplifier has the parameters specified in problem 15. The center frequency is 30 MHz. Other data given:

$Q_{e_1} = 225$ $\qquad Q_{e_2} = 49$ $\qquad R_{ar_p} = 50$ kΩ $\qquad \gamma = 3\gamma_{crit}$

Find: (a) the circuit bandwidth

(b) the L and C of the tuned circuit in the primary

(c) the circuit response at 30.5 MHz.

22. A double tuned transformer-coupled amplifier has the parameters specified in problem 16. The center frequency is 50 MHz. Other data given:

$Q_{e_1} = 256$ $\qquad Q_{e_2} = 64$ $\qquad R_{ar_p} = 40$ kΩ $\qquad \gamma = 2\gamma_{crit}$

Find: (a) the circuit bandwidth

(b) the L and C of the primary tuned circuit

(c) the circuit response at 50.75 MHz.

23. An FET double tuned transformer-coupled amplifier has the following given data:

$r_d = 100 \text{ k}\Omega$ $R_{ar_p} = 600 \text{ k}\Omega$ $Q_o = 250$
$G_m = 5000 \ \mu\text{mhos}$ $a = 1$ $\gamma = 4$
$f_{ar} = 20 \text{ MHz}$

Find: (a) the circuit bandwidth
   (b) the maximum voltage amplification
   (c) the relative response of the circuit at 20.5 MHz.

24. An FET double tuned transformer-coupled amplifier has the following given data:

$r_d = 500 \text{ k}\Omega$ $R_{ar_p} = 100 \text{ k}\Omega$ $a = 1$
$G_m = 2500 \ \mu\text{mhos}$ $f_{ar} = 30 \text{ MHz}$ $\gamma = 3$
$Q_1 = 200$

Find: (a) the maximum voltage amplification
   (b) the circuit bandwidth
   (c) the response of the circuit at 30.1 MHz.

25. An FET double tuned transformer-coupled amplifier has the following given data:

$Q_{e_1} = 300$ $G_m = 3000 \ \mu\text{mhos}$ $\gamma = 3$
$Q_{e_2} = 27$ $f_{ar} = 50 \text{ MHz}$

Find: (a) the circuit bandwidth
   (b) the response of the circuit at 50.75 MHz.

26. An FET double tuned transformer-coupled amplifier has the following given data:

$Q_{e1} = 270$ $G_m = 300 \ \mu\text{mhos}$ $f_{ar} = 50 \text{ MHz}$
$Q_{e2} = 30$ $r_d = 200 \text{ k}\Omega$ $L_1 = 5 \ \mu\text{H}$ $\gamma = 3$

Find: (a) Re
   (b) the circuit bandwidth
   (c) the response of the circuit at 50.9 MHz.

five

RECTIFIERS AND FILTERS

INTRODUCTION

The purpose of a power supply is to convert the Edison line input to a unidirectional output. The technique involved utilizes the process of rectification to convert an ac signal input to a pulse containing dc and harmonic components. The harmonic components are then removed by proper usage of filters.

SINGLE PHASE HALF-WAVE RECTIFIERS

The block diagram of a rectifier circuit is shown in Fig. 5-1(a) and the basic circuit in Fig. 5-1(b).

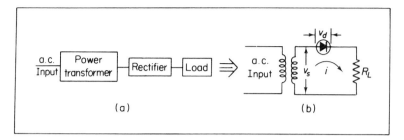

Figure 5-1   *Single Phase Half-Wave Rectifier Circuit*

theory   The primary winding of the power transformer is connected to the ac input. The secondary of the transformer is used to either increase the voltage or current depending on the design requirements. One lead of the secondary transformer is connected to the rectifier and the other lead is connected to the load. The load resistor is connected to the cathode of the rectifier and the system is considered a series circuit.

The induced voltage v_s is directly applied to both the diode and the load. When the application of system voltage is positive to the anode of the diode, current i begins to flow. When the system voltage is negative to the anode of the diode, current ceases to flow. There is a small reverse current that flows, but this current is assumed negligible and the diode is considered nonconducting in this region.

The output and input current and voltage waveforms are shown in Fig. 5-2.

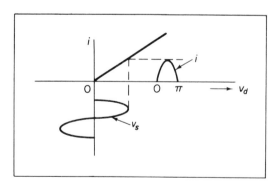

Figure 5-2   *Current Pulses Obtained in Single Phase Half-Wave Rectifier*

Since the current flow is in one direction only, the polarity of voltage across the load resistor must be positive at the cathode end with respect to the other end of the resistor. Also, since the dc current flows through the load, the rectifier and the transformer, the molecules within the iron core of the transformer tend to align themselves in one direction. Consequently, the resultant effect is called *dc core saturation*, with a resultant decrease in the transformer efficiency.

analysis   The equivalent circuit of the single phase half-wave rectifier circuit is shown in Fig. 5-3. Consider v_s to be sinusoidal in nature and equal to:

$$v_s = V_s \sin \omega t$$

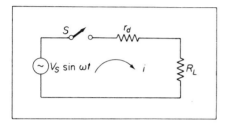

Figure 5-3   *Equivalent Circuit of Single Phase Rectifier*

Note that on the positive half cycle, switch S is closed and current flows, whereas on the negative half cycle, S is open and current flow stops. The waveforms of input voltage and load current are illustrated in Fig. 5-4.

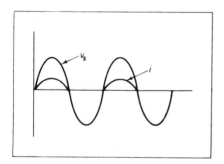

Figure 5-4   *Waveforms in a Single Phase Half-Wave Rectifier Circuit*

The current pulse, i, is defined by:

$$i = I_m \sin \omega t \Big|_0^{\pi}$$

The harmonic components of the current pulse can be expressed by a Fourier series as:

$$i = \frac{I_m}{\pi} + \frac{I_m}{2} \sin \omega t - \frac{2I_m}{3\pi} \cos 2\omega t - \frac{2I_m}{15\pi} \cos 4\omega t \dots$$

Note that the value of I_m is specified by:

$$I_m = \frac{V_s}{r_d + R_L}$$

Examination of the Fourier series indicates that the average or dc term is defined by:

$$I_{dc} = \frac{I_m}{\pi}$$

The total or root mean square value of the current pulse is given by the relationship

$$I_{rms} = \frac{I_m}{2}$$

The form factor F of the circuit is defined as the ratio of the rms value of current to the dc value of current. Thus,

$$F = \frac{I_{rms}}{I_{dc}}$$

For the single phase half-wave rectifier circuit the form factor F is:

$$F = \frac{.5\,I_m}{.318\,I_m} = 1.57$$

In practical power supplies, any variations about the dc level are undesirable. The amount of ripple or ac variation compared to the dc is a measure of the purity of the power supply output and is defined by a *ripple factor*. The value of the ripple factor, symbolized by r, is determined by:

$$r = \frac{I_{rms_{ac}}}{I_{dc}}$$

where:

$I_{rms_{ac}}$ = *rms* value of the ac components only.
I_{dc} = average value of the current pulse.

The value of $I_{rms_{ac}}$ can be determined from the total rms value of the current pulse as:

$$I_{rms_{ac}} = \sqrt{I_{rms}^2 - I_{dc}^2}$$

Removing I_{dc} from the radical results in:

$$I_{rms_{ac}} = I_{dc}\sqrt{\frac{I_{rms}^2}{I_{dc}^2} - 1}$$

or

$$\frac{I_{rms_{ac}}}{I_{dc}} = r = \sqrt{F^2 - 1}$$

The advantage of using this relationship lies in the ease of power supply measurement. Thus, voltage output measurements using dc and ac scales permit rapid evaluation of the ripple factor. Consequently, the ripple factor for the single phase half-wave rectifier circuit is equal to:

$$r = \sqrt{(1.57)^2 - 1} = 1.21$$

efficiency　The efficiency of rectification is defined by the ratio of the output dc power to the total amount of input power supplied to the circuit. The input power is given by:

$$P_i = I_{rms}^2 (r_d + R_L)$$

The output power is given by:

$$P_o = I_{dc}^2 R_L$$

The efficency denoted by the symbol η is evaluated by:

$$\eta = \frac{P_o}{P_i}$$

In terms of percentages, the value of circuit efficiency becomes equal to:

$$\%\eta = \frac{I_{dc}^2 R_L}{I_{rms}^2(r_d + R_L)} 100\ \%$$

Substituting and simplifying results in:

$$\%\eta = \frac{40.6\ \%}{1 + \dfrac{r_d}{R_L}}$$

It is evident that the maximum possible efficiency is 40.6%. For each rectifier circuit, there is a maximum voltage that the diode must withstand. This potential is called the *peak inverse voltage* (PIV) because it occurs during that portion of the cycle when the diode is nonconducting. For the single phase half-wave rectifier the PIV is E_m.

An illustrative problem will demonstrate the theory.

sample problem

A single phase half-wave rectifier circuit supplies power to a 1500 Ω load. The input signal voltage to the rectifier is 300 V rms. The diode resistance is 300 Ω.

Find: (a) I_{dc} (b) ripple voltage (c) efficiency

Solution:

Step 1: Calculate V_s

$$V_s = 1.414V_{rms} = 424.2V$$

Step 2: Calculate I_m

$$I_m = \frac{V_s}{r_d + R_L} = \frac{424.2}{300 + 1500}$$

$$I_m = 235 \text{ mA}$$

Step 3: Calculate I_{dc}

$$I_{dc} = \frac{I_m}{\pi} = \frac{235 \text{ mA}}{\pi}$$

$$I_{dc} = 75 \text{ mA}$$

Step 4: Calculate ripple voltage
The ripple factor is 1.21. The dc voltage across the load is $I_{dc} R_L$, which is equal to 112.5 V.

$$V_{rms_{ac}} = rV_{dc} = 1.21 (112.5)$$

$$V_{rms_{ac}} = 136 \text{ V}$$

Step 5: Calculate circuit efficiency

$$\%\eta = \frac{40.6\%}{1 + \frac{r_d}{R_L}} = \frac{40.6\%}{1 + .2}$$

$$\%\eta = 33.8 \%$$

SINGLE PHASE FULL-WAVE RECTIFIER CIRCUIT

A method of increasing the efficiency of rectification is to supply the load with current during the half cycle when one diode is nonconducting. A circuit called the single phase full-wave rectifier circuit, shown in Fig. 5-5, accomplishes this function.

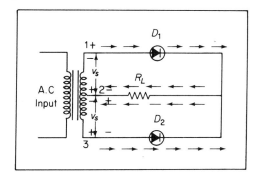

Figure 5-5   *Single Phase Full-Wave Rectifier Circuit*

The two halves of the secondary winding 1-2 and 2-3 is center tapped as shown. The load R_L is connected between the cathodes of D_1 and D_2 and the center tap.

theory   Assume both diodes are conducting equal amounts of current. Consider the application of v_s to diode D_1. During the positive half cycle, the anode of D_1 is positive with respect to K and current flows. The direction of current flow is indicated on the diagram of Fig. 5-5 by the dotted line. During this half cycle, the diode D_2 is nonconducting.

When the input voltage reverses. D_1 ceases conduction and D_2 starts to conduct. The current flow through R_L during this period is shown by the dotted line. It is evident that the current flows through R_L in the same direction. Consequently, the two diodes alternate in synchronous conduction with the input signal acting as a switch. Thus, D_1 is on for a half cycle and D_2 is off; then D_1 is off for a half cycle and D_2 is on. Because there are two pulses of current flowing through the load for each input cycle to the primary of the transformer, the single phase full-wave rectifier circuit utilizes the power transformer in an efficient manner.

analysis   The equivalent circuit of the single phase full-wave rectifier circuit is shown in Fig. 5-6. The circuit waveforms are shown in Fig. 5-7.

The total current that flows through R_L can be designated by i_b and is equal to the sum of the two current pulses i_{b_1} and i_{b_2}. The total current i_b can also be expressed by the Fourier series as:

$$i_b = \frac{2I_m}{\pi} - \frac{4I_m}{3\pi} \cos 2\omega t - \frac{4I_m}{15\pi} \cos 4\omega t - \cdots$$

where:

$$I_m = \frac{V_s}{r_d + R_L}$$

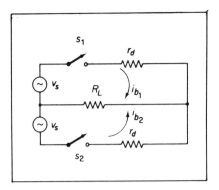

Figure 5-6   *Single Phase Full-Wave Rectifier Circuit*

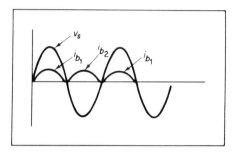

Figure 5-7   *Circuit Waveforms for the Single Phase Full-Wave Rectifier Circuit*

The average or dc value of the load current is given by:

$$I_{dc} = \frac{2I_m}{\pi}$$

The total root mean square value of the current pulse is:

$$I_{rms} = .707I_m$$

The form factor for this circuit is:

$$F = \frac{I_{rms}}{I_{dc}}$$

Substituting and solving for the form factor in a single phase full-wave rectifier circuit results in

$$F = 1.11$$

The ripple factor has the given definition. Thus,

$$r = \sqrt{F^2 - 1}$$

If the value for the form factor is substituted into the ripple factor equation, the resultant value for r is:

$$r = 0.482$$

The efficiency of rectification is given by the relationship

$$\%\eta = \frac{I_{dc}^2 R_L}{I_{rms}^2 (r_d + R_L)} \, 100 \, \%$$

or

$$\%\eta = \frac{81.2 \,\%}{1 + \dfrac{r_d}{R_L}}$$

It is evident that the maximum possible efficiency for the single phase full-wave rectifier circuit is 81.2%. Another rating considered important by the design engineer is the peak inverse voltage rating of each diode. At any instant of time, diode D_1 is conducting and diode D_2 is nonconducting. If Kirchoff's voltage law is applied to the outer loop and assuming D_1 voltage drop is negligible, then the value of PIV across D_2 is $2\,V_s$. Note that the result obtained is independent of the type of load used. For example, the load may be purely resistive or a combination of resistive and reactive elements.

An illustrative problem will demonstrate the theory.

sample problem

A single phase full-wave rectifier circit supplies power to a 1500 Ω load. The ac voltage applied to the diode is $300 - 0 - 300$ V *rms*. The diode resistance is 300 Ω.

Find: (a) I_{dc} (b) the ripple voltage (c) efficiency

Solution:

Step 1: Calculate V_s

$$V_s = 1.414 \, V_{rms} = 1.414 \, (300)$$

$$V_s = 424.2 \text{ V}$$

Step 2: Calculate I_m

$$I_m = \frac{V_s}{r_d + R_L} = \frac{424.2}{300 + 1500}$$

$$I_m = 235 \text{ mA}$$

Step 3: Calculate I_{dc}

$$I_{dc} = \frac{2I_m}{\pi} = \frac{2 \times 235 \text{ mA}}{\pi}$$

$$I_{dc} = 150 \text{ mA}$$

Step 4: Calculate the ripple voltage. The ripple factor is 0.482. The dc voltage is given by $I_{dc} R_L$ or 225 V. Thus,

$$V_{rms_{ac}} = rV_{dc} = 0.482 \times 225$$

$$V_{rms_{ac}} = 108 \text{ V } rms$$

Step 5: Calculate the circuit efficiency

$$\%\eta = \frac{81.2\%}{1 + .2}$$

$$\%\eta = 67.6\%$$

Step 6: Construct the circuit

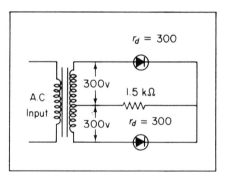

SINGLE PHASE FULL-WAVE BRIDGE CIRCUIT

Other types of rectifier circuits are useful in certain applications. Some of these are the single phase full-wave bridge rectifier circuit, voltage multipliers and so forth. The basic circuit of a single phase full-wave bridge rectifier circuit is shown in Fig. 5-8.

Consider the instantaneous voltage across the input terminals of the power transformer to be instantly positive to the left and negative to the right. Note that during this period, diodes D_1 and D_4 are conducting and the current flows through the load with the polarity shown. The conduction path is shown by the solid arrows. During the next half cycle, the transformer polarity is reversed and diodes D_2 and D_3 are conducting. The load current is in the same direction as before and is shown by the dotted arrows.

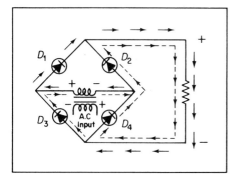

Figure 5-8 *Single Phase Full-Wave Bridge Rectifier Circuit*

Some of the advantages of this circuit are:

1. No center tap required on transformers.
2. Smaller transformers can be used.
3. Suitable for high voltage applications.
4. The PIV rating per diode is less. For the single phase full-wave rectifier the PIV for any diode is $2 V_s$. For the single phase full-wave bridge rectifier circuit, the PIV per diode is V_s.

The disadvantages of the bridge rectifier circuit are:

1. Poorer regulation.
2. More expensive; uses additional diodes and sockets.

The efficiency of the bridge rectifier circuit is almost identical with that of the single phase full-wave rectifier circuit. Thus, the efficiency of rectification is:

$$\%\pi = \frac{81.2\%}{1 + \dfrac{2r_d}{R_L}}$$

The essential difference between the bridge rectifier and the simple rectifier circuit is based on the PIV rating of the diodes used.

THE SERIES INDUCTOR FILTER

In general, a power supply must supply a ripple-free undirectional voltage and current to a load. Consequently, the harmonic components contained in the rectifier output pulses must be drastically reduced or eliminated. A network approach utilizes a filter between the rectifier and load.

An inductor inserted between the rectifier and the load presents a high impedance to the alternating current components of the current pulse. The action of this filter utilizes the fundamental properties of an inductor— that is, to oppose any changes in current. Therefore, any sudden changes that may occur are smoothed out. Note that magnetic energy is stored in the inductor when the current tends to rise above the average value and is returned to the circuit when the current falls below the average value.

The basic circuit of a single phase half-wave rectifier with an inductor filter inserted between the rectifier and load is shown in Fig. 5-9.

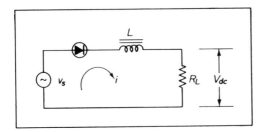

Figure 5-9 *Half-Wave Rectifier with Choke Filter*

analysis The equation for the circuit is:

$$v_s = L\frac{\Delta i}{\Delta t} + R_L i$$

where

$$v_s = V_s \sin \omega t$$

The complete solution contains two parts, a transient solution and a steady state solution. The transient solution is:

$$i_{\text{trans}} = \frac{V_s}{Z} e^{-R_L t/L} \sin \theta$$

where

$$Z = \sqrt{R_L^2 + \omega^2 L^2}$$

$$\theta = \tan^{-1} \frac{\omega L}{R_L}$$

The steady state solution is:

$$i_{ss} = \frac{V_s}{Z} \sin (\omega t - \theta)$$

The total current combines the transient and the steady state solutions. Thus,

$$i = i_{\text{trans}} + i_{ss}$$

or

$$i = \frac{V_s}{Z}[\sin(\omega t - \theta) + e^{-R_L t/L} \sin \theta]$$

The only condition required to maintain the validity of the solution is that current i is equal to zero when ωt is equal to zero. For different values of L, the current waveforms will vary accordingly. This variation is shown in Fig. 5-10.

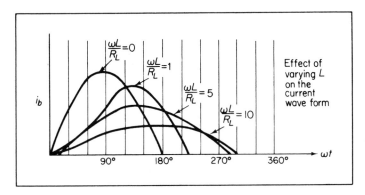

Figure 5-10 *Effect of Varying L on the Current Waveform*

Because of the discontinuous conduction the single phase half-wave rectifier circuit produces a low dc output voltage and is rarely used with an inductor filter.

FULL-WAVE RECTIFIER WITH SERIES INDUCTOR FILTER

A single phase full-wave rectifier circuit using a series inductor filter is shown in Fig. 5-11.

theory The rectifier output pulse is applied to the LR output circuit. The current variation tends to level off due to the action of the inductor. Consequently, the waveforms of voltage and current that appear in the circuit are shown in Fig. 5-12.

analysis The analysis is simplified because of the continuous flow of current. The mathematical expression for the voltage applied to the input of

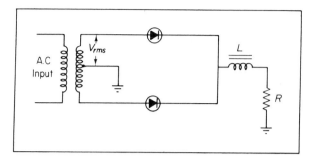

Figure 5-11 *Full-Wave Rectifier Circuit with Series Inductor Filter*

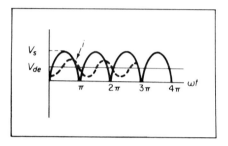

Figure 5-12 *Waveforms in a Full-Wave with Series Inductor Filter*

the filter is given by

$$v_i = \frac{2V_s}{\pi} - \frac{4V_s}{\pi}\left(\frac{1}{3}\cos 2\omega t + \frac{1}{15}\cos 4\omega t + \frac{1}{35}\cos 6\omega t + \cdots\right)$$

Investigation of the dc and harmonic components applied to the filter leads to the following conclusions:

1. $V_{dc} = \dfrac{2V_s}{\pi}$

2. Fundamental peak magnitude $= \dfrac{4V_s}{3\pi}$

If the circuit uses a frequency of 60 Hz input to the rectifiers, the fundamental frequency applied to the filter is 120 Hz.

3. Second harmonic peak magnitude $= \dfrac{4V_s}{15\pi}$

The frequency of the second harmonic input to the filter is 240 Hz.

4. Third harmonic peak magnitude $= \dfrac{4V_s}{35\pi}$

The frequency of the third harmonic input to the filter is 360 Hz. The magnitude of the second harmonic is 20 percent of the fundamental magnitude. The magnitude of the third harmonic is approximately 8.6 percent of the fundamental.

It is evident that small error will be introduced if all the harmonics are neglected and the input to the filter assumed as:

$$v_i = \frac{2V_s}{\pi} - \frac{4V_s}{3\pi} \cos 2\omega t$$

In many cases, it is necessary to find the ripple output voltage.
Procedure:

1. The first step is to establish the rms value of the ac components to the input of the filter. In this case, the value is determined by:

$$V_{rms_{ac}} = \frac{4V_s}{3\pi\sqrt{2}}$$

2. The value of circuit impedance must be evaluated. Thus,

$$Z = \sqrt{R_L^2 + \omega^2 L^2}$$

The frequency used is the input fundamental to the filter.

3. The output ripple voltage is then determined using a voltage divider action. The output voltage will be primed to distinguish it from the input signal voltage. Thus,

$$V'_{rms_{ad}} = V_{rms_{ac}} \frac{R_L}{Z}$$

An illustrative problem will demonstrate the theory.

sample problem

A single phase full-wave rectifier circuit uses a series inductor between rectifiers and load. The load resistor is $2\,k\Omega$. The transformer input is 400 V $-\ 0\ -$ 400 V *rms* and the value of L is 10 H. Find:

(a) the dc output voltage
(b) the rms_{ac} value of the load voltage

Solution:

Step 1: Calculate V_{dc}

$$V_{dc} = \frac{2V_s}{\pi} = \frac{2 \times 565.6}{\pi}$$

$$V_{dc} = 360 \text{ V}$$

Step 2:   Calculate the input rms_{ac} voltage to the filter

$$V_{rms_{ac}} = \frac{4V_s}{3\pi\sqrt{2}} = \frac{4 \times 565.6}{3\pi \times 1.414}$$

$$V_{rms_{ac}} = 170 \text{ V}$$

Step 3:   Calculate Z

$$Z = \sqrt{(2 \times 10^3)^2 + (2\pi \times 120 \times 10)^2}$$

$$Z = \sqrt{(4 + 57)\,10^6}$$

$$Z = 7.81 \text{ k}\Omega$$

Step 4:   Calculate $V'_{rms_{ac}}$

$$V'_{rms_{ac}} = V_{rms_{ac}}\frac{R}{Z}$$

$$= 170\left(\frac{2}{7.81}\right)$$

$$= 43.5 \text{ V}$$

CAPACITANCE FILTER

Many practical systems use a capacitor shunted across the load to achieve filtering action. This type of filter is known as a capacitance input filter. The basic circuit of a single phase half-wave rectifier circuit with filter is shown in Fig. 5-13.

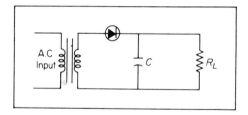

Figure 5-13   *Half-Wave Rectifier with Capacitor Filter*

The filter circuit depends for its operation on the property of a capacitor to charge, or store energy, during conduction and to discharge, or deliver energy, during the nonconduction cycle. The reactance of the capacitor is assumed much smaller than the load resistance. Consequently, the application of the first positive half cycle to the filter capacitor causes the capacitor to charge to the peak of the input pulse. During the negative half cycle the rectifier is nonconducting. Capacitor C must discharge through R_L during this period of time. Consequently, the voltage across C rises and falls dependent on the $R_L C$ time constant. The waveforms inherent in the filter system are shown in Fig. 5-14.

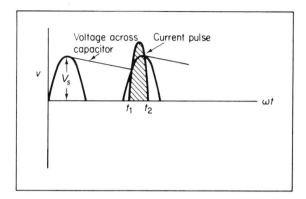

Figure 5-14   *Waveforms Associated with Capacitor Filter*

Because of the charge and discharge of the capacitor, the input waveforms to the filter must be analyzed to determine the average and harmonic components. The actual and approximate waveforms are shown in Fig. 5-15.

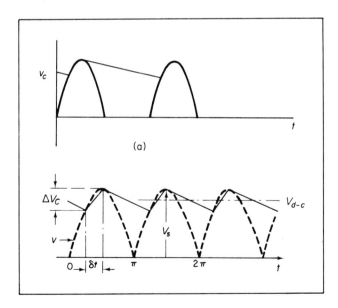

Figure 5-15   *Actual and Approximate Waveforms in a Capacitor Input Filter:*
(a) Actual Waveform (b) Approximate Input Wave

The average value of the output voltage is given by the relationship

$$V_{dc} = V_s - \frac{\Delta V_c}{2}$$

The *rms* value of the ac components only can be determined by:

$$V_{rms_{ac}} = \frac{\Delta V_c}{2\sqrt{3}}$$

The assumption must be made that the charge of the capacitor is equal to the discharge of the capacitor. Consequently, the charge of the capacitor is:

$$(\Delta V_c)C = Q_{charge}$$

The discharge of the capacitor is given by:

$$I_{dc}T = Q_{discharge}$$

where T is the period of time between successive capacitor chargings. Thus,

$$I_{dc}T = (\Delta V_c)C$$

The ripple factor is given by the ratio of the *rms* value of the ac components to the dc voltage. Thus,

$$\text{ripple factor } (r) = \frac{V_{rms_{ac}}}{V_{dc}}$$

Substituting for $V_{rms_{ac}}$ the value of $\dfrac{\Delta V_c}{2\sqrt{3}}$ and for V_{dc} the value of $I_{dc}R_L$ and with some algebraic manipulation, the result is:

$$r = \frac{T}{2\sqrt{3}\,R_L C}$$

An illustrative problem will demonstrate the theory.

sample problem

A single phase half-wave rectifier circuit uses a capacitor filter across the resistive load. The following data are given:

$$V_{dc} = 350 \text{ V} \qquad I_{dc} = 200 \text{ mA} \qquad C = 20 \text{ }\mu\text{F}$$

Find: (a) $V_{rms_{ac}}$ (b) $V'_{rms_{ac}}$ (c) ripple factor

Solution:

    Step 1:   Calculate ΔV_s

$$(\Delta V_c)C = I_{dc}\ T$$

$$\Delta V_c = \frac{.2 \times \frac{1}{60}}{20 \times 10^{-6}}$$

$$\Delta V_c = 167\ \text{V}$$

    Step 2:   Calculate $V_{rms_{ac}}$

$$V_{rms_{ac}} = \frac{\Delta V_c}{2\sqrt{3}}$$

$$V_{rms_{ac}} = 48.3\ \text{V}$$

    Step 3:   In this case, $V'_{rms_{ac}}$ is exactly equal to the applied ac voltage. Thus, $V'_{rms_{ac}} = 48.3$ V

    Step 4:   Calculate the ripple factor

$$r = \frac{T}{2\sqrt{3}\ R_L C}$$

$$r = \frac{0.167}{2 \times 1.732 \times 1.75 \times 10^3 \times 20 \times 10^{-6}}$$

$$r = .1375$$

It is evident that increasing R_L or C reduces the ripple factor. Large capacitors, therefore, are used to keep the ripple voltage low. In general, the capacitor filter is characterized by a higher dc load voltage with a lower ripple voltage than the series inductor filter. On the other hand, the series inductor filter has a superior voltage regulation with a consequent reduction in current peaks.

LC FILTER

The series inductor and the shunt capacitor filters may be combined for improved filtering action. The combination forms the *choke input filter* and is shown in Fig. 5-16.

This method of filtering affords a greater reduction in ripple voltage than either of the previous filters. The series inductor presents an extremely high input impedance to the harmonic components of the input pulse. The shunt capacitor presents a comparatively low impedance to the ripple components, shorting them out to ground. Consequently, the output dc voltage will have very little output ripple or ac variation.

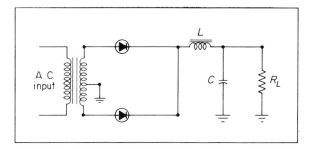

Figure 5-16 *Choke Input Filter*

analysis   A filter of this type is seldom used with a single phase half-wave rectifier because most circuits cannot tolerate discontinuous current flow. Consequently, this analysis will assume a single phase full-wave rectifier supplying a choke input filter.

The dc and harmonic components representing the rectifier output pulse are given by:

$$v = \frac{2V_s}{\pi} - \frac{4V_s}{\pi}\left(\frac{1}{3}\cos 2\omega t + \frac{1}{15}\cos 4\omega t + \frac{1}{35}\cos 6\omega t + \cdots\right)$$

The filter elements are usually selected as follows:

1. The series inductor must present a high impedance to the harmonic components.
2. The capacitor must present a low impedance to the harmonic components.

Based on these conditions, the following assumptions can be made:

1. The inductive reactance is much greater than the capacitive reactance at the fundamental frequency input to the filter.

$$X_L \gg X_c$$

2. Since the capacitive reactance of the filter is low, the load resistance is also much greater measured at the fundamental frequency input to the filter. Thus,

$$R_L \gg X_c$$

3. The *rms* value of the fundamental input of the harmonic components will be assumed applied to the choke input filter. This value is:

$$V_{rms_{ac}} = \frac{4V_s}{3\pi\sqrt{2}}$$

It is the purpose of all filters to reduce or eliminate the ripple variation about the dc component. The resultant equivalent circuit is shown in Fig. 5-17.

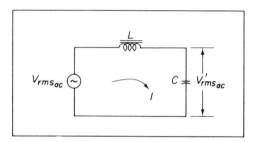

Figure 5-17   *Equivalent Circuit of Choke Input Filter*

Note that the current flow in the circuit is given by:

$$I = \frac{V_{rms_{ac}}}{j(X_L - X_c)}$$

and

$$V'_{rms_{ac}} = I(-jX_c)$$

The magnitude of the output ripple voltage is given by:

$$V'_{rms_{ac}} = V_{rms_{ac}}\frac{X_c}{X_L - X_c}$$

Let $\alpha = \dfrac{X_c}{X_L - X_c}$, then $V'_{rms_{ac}} = \alpha V_{rms_{ac}}$

The input frequency to the filter is usually 120 Hz, assuming Edison line frequency input to the rectifiers. The factor α is defined as the smoothing factor or how much the input ripple voltage has been reduced.

The method of calculating V_{dc} utilizes the following equation:

$$V_{dc} = \frac{2V_s}{\pi} = .9\,V_{rms}$$

The ripple factor at the input to the filter is given by:

$$r = \frac{V_{rms_{ac}}}{V_{dc}} = \frac{\dfrac{4V_{rms}}{3\pi}}{.9V_{rms}}$$

$$r = 0.471$$

The output ripple factor is given by r' and is equal to:

$$r' = \alpha r = 0.471\alpha$$

Note that the ripple factor will be independent of load. An illustrative problem will demonstrate the theory.

sample problem

A single phase full-wave rectifier uses a choke input filter. The following data are given:

$$V_{rms} = 300 \text{ V} \qquad L = 10 \text{ H} \qquad C = 20 \text{ } \mu\text{F}$$

Find: (a) V_{dc} (b) $V_{rms_{ac}}$ (c) $V'_{rms_{ac}}$ (d) r'

Solution:

Step 1:   Calculate V_{dc}

$V_{dc} = .9V_{rms}$

$V_{dc} = 270 \text{ V}$

Step 2:   Calculate $V_{rms_{ac}}$

$$V_{rms_{ac}} = \frac{4V_{rms}}{3\pi}$$

$V_{rms_{ac}} = 127.2 \text{ V}$

Step 3:   Calculate $V_{rms_{ac}}$

$$V'_{rms_{ac}} = \frac{V_{rms_{ac}}}{\omega^2 LC - 1}$$

$$V'_{rms_{ac}} = \frac{127.2}{(2\pi \times 120)^2 \, 10 \times 20 \times 10^{-6} - 1}$$

$V'_{rms_{ac}} = 1.125 \text{ V}$

Step 4:   Calculate r'

$r' = 0.471\alpha = 0.471 \times 8.85 \times 10^{-3}$

$r' = 4.16 \times 10^{-3}$

CRITICAL VALUE OF THE FILTER INPUT COIL

A power supply must provide continuous load power. Consequently, the peak value of the fundamental current must always be less than the output dc current. That is:

$$I_{dc} \geq I_m$$

The possible variation of the fundamental component around the dc level is shown in Fig. 5-18.

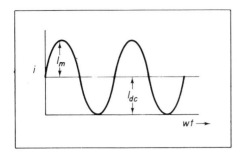

Figure 5-18 *Current Waveforms Through the Choke Input Filter*

The critical value of the inductor required to maintain a continuous flow of load current is evaluated by the following analysis.

analysis The analysis is based on the premise that

$$I_{dc} \geqq I_m$$

Note that

$$I_{dc} = \frac{V_{dc}}{R_L} \quad \text{and} \quad I_m = \frac{4V_s}{3\pi(X_L - X_c)}$$

Since $V_{dc} = \dfrac{2V_s}{\pi}$ and X_L is much greater than X_c, then substituting these values into the first equation yields:

$$\frac{2V_s}{\pi R_L} \geqq \frac{4V_s}{3\pi X_L}$$

The critical value of L is given by:

$$L \geqq \frac{R_L}{3\pi f} \text{ H}$$

If $f = 120$ Hz, then the value of L is:

$$L \geqq \frac{R_L}{1130} \text{ H}$$

The value of critical inductance used various assumptions. Consequently, some error may have been introduced. A practical design equation commonly used is:

$$L \geqq \frac{R_L}{1000} \text{ H}$$

It should be noted that insertion of an inductor less than the critical value will require a new analysis for the circuit. In practical circuits, a series inductor called a swinging choke is utilized. This type of choke has an internal air gap that produces the required choke variation. This choke is designed for critical inductance at rated load current. A typical swinging choke will vary from 4 H at 100 − 150 mA to about 30 H at a few milliamperes.

An illustrative problem will demonstrate the theory.

sample problem

Design a single phase full-wave rectifier circuit with choke input filter to supply a load with 450 V dc and a load current of 150 mA. The output ripple voltage should not exceed three percent of the load voltage. Specify all circuit elements.

Solution:

   Step 1:   Calculate R_L

$$R_L = \frac{450}{.15} = 3 \text{ k}\Omega$$

   Step 2:   Calculate the critical inductance

$$L_c = \frac{R_L}{1000} = \frac{3000}{1000}$$

$$L_c = 3 \text{ H}$$

      The value of 3 H specifies the minimum allowable value of inductance that can be used.

   Step 3:   Calculate the output ripple voltage

$$V_r' = 0.03 \times 450 = 13.5 \text{ V}$$

   Step 4:   Calculate the value of α

$$r' = 0.471 \, \alpha$$

$$.03 = .471 \, \alpha$$

$$\alpha = .0636$$

   Step 5:   Calculate the value of LC

$$\omega^2 LC = \frac{1}{\alpha} + 1 = 16.7$$

$$LC = \frac{16.7}{57 \times 10^4} = 29.3 \times 10^{-6}$$

$$C = 9.77 \, \mu\text{F}$$

MULTIPLE LC SECTION FILTER

The output ripple voltage can be further reduced or attenuated by cascading *LC* sections. A two-section *LC* type filter is shown in Fig. 5-19.

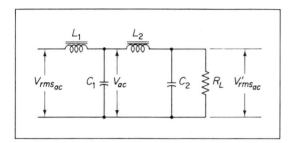

Figure 5-19 *Two-Section LC Filter*

The circuit analysis will assume that in filters the capacitive reactance, X_c, is always much smaller than the inductive reactance, X_L.

analysis The ripple voltage across C_1 assumes that X_{L_2} is extremely large compared to X_{c_1}. Then the second harmonic current passes predominantly through C_1. Thus,

$$V_{ac} = \frac{V_{rms_{ac}}}{\omega^2 L_1 C_1 - 1}$$

or

$$V_{ac} = \alpha_1 V_{rms_{ac}}$$

where

$$\alpha_1 = \frac{1}{\omega^2 L_1 C_1 - 1}$$

The output ripple voltage across C_2 is based on the assumptions that:

1. The input to the second section is V_{ac}.
2. The ohmic impedance of X_{c_2} is much smaller than R_L.

Then,

$$V'_{rms_{ac}} = \frac{V_{ac} \times 1}{\omega^2 L_2 C_2 - 1}$$

or

$$V'_{rms_{ac}} = \alpha_2 V_{ac}$$

where

$$\alpha_2 = \frac{1}{\omega^2 L_2 C_2 - 1}$$

Consequently, the output ripple voltage can be determined by:

$$V'_{rms_{ac}} = \alpha_1 \alpha_2 V_{rms_{ac}}$$

The product $\alpha_1 \alpha_2$ is the total smoothing factor for the two sections. The calculation of ω is made at the input fundamental frequency to the filter. Thus, if Edison line frequency is applied to the rectifiers, the input frequency to the filter is 120 Hz.

An illustrative problem will demonstrate the theory.

sample problem

A filter section is to be connected to a single phase full-wave rectifier. The input voltage to the rectifiers is $300 - 0 - 300$ V *rms*. In stock there are two 20 μF capacitors and two 10 H chokes. Find:

(a) the output ripple voltage when the two chokes are series connected and the two capacitors shunt connected as shown.

(b) the output ripple voltage when the circuit elements are used to form a two-section L filter.

Solution:

Step 1: The input ripple voltage to the filter is:

$$V_{rms_{ac}} = \frac{4V_{rms}}{3\pi}$$

$$V_{rms_{ac}} = 127.2 \text{ V}$$

Step 2: The filter is a single LC section. Thus,

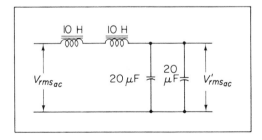

Step 3: Calculate $V'_{rms_{ac}}$

$$V'_{rms_{ac}} = \frac{127.2}{\omega^2 (20)(40) \times 10^{-6} - 1}$$

$$V'_{rms_{ac}} = 280 \text{ mV}$$

Step 4:   Connect the elements of the filter into a two-section LC filter as shown. Calculate $V'_{rms_{ac}}$

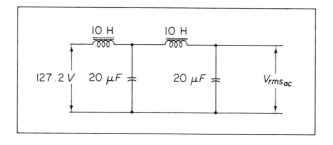

$$V'_{rms_{ac}} = \alpha_1 \alpha_2 V_{rms_{ac}}$$

$$\alpha_1 = \frac{1}{\omega^2 L_1 C_1 - 1}$$

$$\alpha_1 = 8.85 \times 10^{-3} = \alpha_2$$

$$V'_{rms_{ac}} = 78 \times 10^{-6} \times 127.2 \text{ V}$$

$$V'_{rms_{ac}} = 10 \text{ mV}$$

It is evident that use of the same components with proper circuitry can reduce the ripple output voltage considerably. The illustrative problem has demonstrated this statement adequately.

π FILTER

Placing a capacitor across the input terminals of the LC filter forms the π filter or capacitor input filter shown in Fig. 5-20.

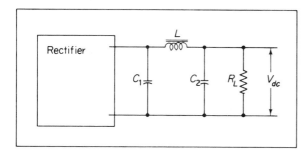

Figure 5-20   π *Section Filter*

theory   The charging and discharging action of the input capacitor C_1 has been previously explained. Capacitor C_1 is charged to the peak value of the input signal voltage. The capacitor C_1 then discharges through L and

R_L until a second pulse is applied to C_1 to charge the capacitor to the peak value of the input pulse. The waveforms across C_1 were shown previously in Fig. 5-15.

analysis   The analysis is identical to that of the capacitance filter. The dc output voltage is given by the relationship:

$$V_{dc} = V_s - \frac{\Delta V_c}{2}$$

The input capacitor changes the input waveform. The method for determining the value of (ΔV_c) is:

$$(\Delta V_c)C_1 = I_{dc}T$$

The LC_2 section filter is used to smooth out the ripple variation applied to the input. The ripple voltage applied to the input of the filter is given by:

$$V_{rms_{ac}} = \frac{\Delta V_c}{2\sqrt{3}}$$

The output ripple voltage is evaluated by:

$$V'_{rms_{ac}} = \frac{V_{rms_{ac}}}{\omega^2 LC_2 - 1}$$

or

$$V'_{rms_{ac}} = \alpha V_{rms_{ac}}$$

where

$$\alpha = \frac{1}{\omega^2 LC_2 - 1}$$

In the capacitor input filter, the ripple output voltage increases with load current. The input capacitance provides a higher output dc voltage from a given input than is possible with the choke input filter and produces a smaller ripple output. The regulation of the capacitance input filter is much poorer than the LC filter.

An illustrative problem will demonstrate the theory.

sample problem

A single plase full-wave rectifier circuit is connected to a π section filter as shown in the figure below. Find:

(a) V_{dc} (b) $V'_{rms_{ac}}$

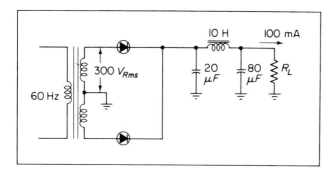

Solution:

Step 1:   Calculate ΔV_c

$$\Delta V_c = \frac{I_{dc} T}{C_1}$$

$$\Delta V_c = \frac{.1 \times \left(\frac{1}{120}\right)}{20 \times 10^{-6}}$$

$$\Delta V_c = 41.7 \text{ V}$$

Step 2:   Calculate V_{sm}

$$V_{sm} = \sqrt{2}\, V_{rsm}$$

$$V_{sm} = 1.414(300)$$

$$V_{sm} = 424.2 \text{ V}$$

Step 3:   Calculate V_{dc}

$$V_{dc} = V_{sm} - \frac{\Delta V_c}{2}$$

$$V_{dc} = 424.2 - \frac{41.7}{2}$$

$$V_{dc} = 403.8 \text{ V}$$

Step 4:   Calculate the value of $V_{rms_{ac}}$

$$V_{rms_{ac}} = \frac{\Delta V_c}{2\sqrt{3}}$$

$$V_{rms_{ac}} = \frac{41.7}{2 \times 1.732}$$

$$V_{rms_{ac}} = 11.74 \text{ V}$$

Step 5: Calculate the smoothing factor

$$\alpha = \frac{1}{\omega^2 L C_2 - 1}$$

$$\alpha = \frac{1}{4\pi^2 (120)^2 10 \times 80 \times 10^{-6} - 1}$$

$$\alpha = 2.2 \times 10^{-3}$$

Step 6: Calculate $V'_{rms_{ac}}$

$$V'_{rms_{ac}} = \alpha V_{rms_{ac}}$$

$$V'_{rms_{ac}} = 2.2 \times 10^{-3} \times 11.74$$

$$V'_{rms_{ac}} = 25.8 \text{ mV}$$

Table 5-1 summarizes all filter circuit operations.

TABLE 5-1 *FILTER CIRCUIT OPERATIONS*

filter	V_{dc}	I_{dc}	$V_{rms_{ac}}$	$V'_{rms_{ac}}$	features
inductor				$V_{rms_{ac}} \dfrac{R_L}{z}$	used with 1ϕ-F.W. rectifiers only
	$\dfrac{2V_{sm}}{\pi}$	$\dfrac{V_{dc}}{R_L}$	$\dfrac{4V_{sm}}{3\pi\sqrt{2}}$	where $z\sqrt{R_L^2 X_L^2}$	
used with F.W.					
capacitor	$V_{sm} - \dfrac{\Delta V_c}{2}$ or	$\dfrac{(\Delta V_c)C}{T}$ or $\dfrac{V_{dc}}{R_L}$	$\dfrac{\Delta V_c}{2\sqrt{3}}$	$\dfrac{\Delta V_c}{2\sqrt{3}}$	can be used with either 1ϕ-HW or F.W. rectifiers
choke input					used with 1ϕ-F.W. rectifiers only
	$\dfrac{2V_{sm}}{\pi}$	$\dfrac{V_{dc}}{R_L}$	$\dfrac{4V_{sm}}{3\pi\sqrt{2}}$	$\dfrac{V_{rms_{ac}}}{\omega^2 LC - 1}$	
π type	$V_{sm} - \dfrac{\Delta V_c}{2}$	$\dfrac{(\Delta V_c)C_1}{T}$	$\dfrac{\Delta V_c}{2\sqrt{3}}$	$\dfrac{V_{rms_{ac}}}{\omega^2 LC_2 - 1}$	can be used with either 1ϕ-HW or F.W. rectifiers

TROUBLESHOOTING ANALYSIS

Power supply troubles are generally categorized by either no dc output or incorrect dc output. When a system fails to perform in more than one of its normal functions the power supply circuit is probably the culprit. The situation of no dc output voltage can occur if there is no output from the rectifiers, no ac input to the power supply or an open filter choke. Figure

5-21 illustrates the proper procedure of using a voltmeter to isolate the defective component. Table 5-2 shows which voltages are missing, which are present and the possible causes.

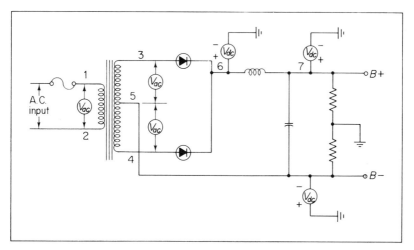

Figure 5-21 *Troubleshooting a Power Supply*

TABLE 5-2 *POWER SUPPLY TROUBLES*

Voltage missing	Voltage present	Probable cause
all	none	blown fuse shorted primary winding
5, 6, 7 to ground	1–2, 3–5, 4–5	rectifier diodes
3–5, low 5, 6, 7 to ground	1–2, 4–5	half of secondary winding either open or shorted
5, 7 to ground	1–2, 3–5, 4–5 6 to ground	open filter choke

A low dc output voltage can occur in the event of:

1. Decreased ac voltage
2. Increased resistance in the diode rectifiers
3. Partial short across the load.

If the power supply system utilizes a capacitor input filter, then if the first filter opens, the output dc voltage will drop. If the first filter capacitor shorts, the output voltage goes to zero.

VOLTAGE MULTIPLIERS

In many circuit applications, it may be necessary to obtain high voltages with low currents. One method is shown in the single phase half-wave doubler circuit shown in Fig. 5-22.

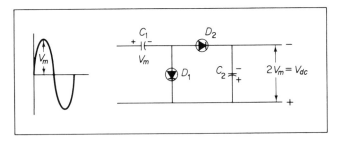

Figure 5-22 *Voltage Doubler Circuit*

The circuit must be assumed operating under steady-state conditions. On the positive half cycle D_1 permits capacitor C_1 to change to the peak of the signal input (V_m). During the negative half cycle the voltage across C_1 adds to the input and discharges completely through D_2 charging C_2 to the sum of the peak signal input plus the capacitor peak charge voltage, or $2V_m$. This process is repetitive and will perform in a satisfactory manner provided the current requirement is small.

By connecting additional diodes and capacitors to the doubler the circuit can be made to triple, quadruple or increase to any multiple of the signal peak input voltage. A circuit that develops three and four times the amplitude of the input signal voltage is shown in Fig. 5-23.

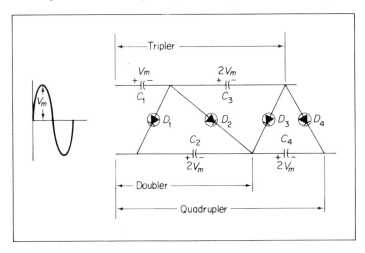

Figure 5-23 *Voltage Tripler and Quadrupler*

theory During the first positive half cycle capacitor C_1 charges as diode D_1 conducts. During the negative half cycle C_2 is charged to twice the peak value of the input signal as diode D_2 conducts. During the second half positive cycle diode D_3 conducts and capacitor C_3 charges to twice the peak value of the input voltage. This charge and discharge sequence is repetitive as additional diodes and capacitors are added. The voltage regulation of voltage multipliers is extremely poor. Generally, these circuits are used where both the supply voltage and load are maintained constant. In addition, it should be noted that the ripple voltage of the multiplier occurs at a frequency of twice the input frequency.

problems

1. A half-wave rectifier uses a diode with r_d equal to one ohm. If the applied voltage is 36 V *rms* and the load is 19 Ω, find: (a) I_{dc} (b) $V_{rms_{ac}}$ (c) η.

2. A half-wave rectifier uses a diode with $r_d = 250$ Ω. If the applied voltage is 300 V *rms*, and the load is 3 kΩ, find: (a) I_{dc} (b) $V_{rms_{ac}}$ (c) η.

3. A full-wave rectifier uses a diode with $r_d = 5$ Ω. If the applied voltage is $50 - 0 - 50$ V *rms* and the load is 45 Ω, find: (a) I_{dc} (b) $V_{rms_{ac}}$ (c) η.

4. A full-wave rectifier uses a diode with $r_d = 300$ Ω. If the applied voltage is $300 - 0 - 300$ V *rms* and the load is 3 kΩ, find: (a) I_{dc} (b) $V_{rms_{ac}}$ (c) η.

5. A full-wave rectifier uses a diode with $r_d = 250$ Ω. If the applied voltage is 800 V *rms* from end to end of the transformer, and the load is 2.5 kΩ, find: (a) I_{dc} (b) $V_{rms_{ac}}$ (c) η.

6. A full-wave rectifier uses a 10 H series inductor filter. The transformer input is $300 - 0 - 300$ V *rms* and the load resistance is 2 kΩ. Find: (a) I_{dc} (b) $V'_{rms_{ac}}$.

7. A full-wave rectifier uses a 15 H series inductor filter. The transformer input is $400 - 0 - 400$ V *rms* and the load resistance is 2.5 kΩ..Find: (a) I_{dc} (b) $V'_{rms_{ac}}$.

8. A half-wave rectifier uses a 20 μF capacitor across a load. The dc load current is 100 mA and the transformer input is 300 V *rms*. Find: (a) I_{dc} (b) $V'_{rms_{ac}}$.

9. A full-wave rectifier uses a 40 μF capacitor across the load. The load current is 100 mA and the transformer input is $400 - 0 - 400$ V *rms*. Find: (a) V_{dc} (b) $V'_{rms_{ac}}$.

10. An *LC* section filter is connected to a full-wave rectifier circuit. The transformer input is $300 - 0 - 300$ V *rms*, $L = 10$ H, $C = 40$ μF, $R_L = 2$ kΩ. Find: (a) V_{dc} (b) I_{dc} (c) $V'_{rms_{ac}}$.

11. A full-wave rectifier circuit using an *LC* section filter is supplying power to a 3 kΩ load. The transformer input is $350 - 0 - 350$ V *rms*, $L = 10$ H, $C = 80$ μF. Find: (a) V_{dc} (b) I_{dc} (c) $V'_{rms_{ac}}$.

The following circuit is given for problems 12 through 16.

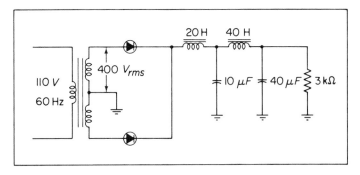

12. For the circuit shown, determine V_{dc} and $V'_{rms_{ac}}$.

13. The 40 H choke is shorted. Find V_{dc} and $V'_{rms_{ac}}$.

14. The 10 μF capacitor is open circuited. Find: V_{dc} and $V'_{rms_{ac}}$.

15. The 10 μF capacitor is open circuited. The 40 H choke is shorted. Find: (a) V_{dc} (b) $V'_{rms_{ac}}$.

16. The 40 μF capacitor is open circuited. Find V_{dc} and $V'_{rms_{ac}}$.

The following circuit is given for problems 17 to 27.

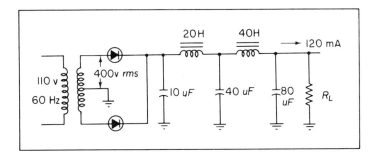

17. Determine V_{dc} and $V'_{rms_{ac}}$.

18. The 20 H choke is shorted. Determine V_{dc} and $V'_{rms_{ac}}$.

19. The 40 μF capacitor is open. Determine V_{dc} and $V'_{rms_{ac}}$.

20. The 40 μF and the 80 μF capacitors are open. Determine V_{dc} and $V'_{rms_{ac}}$.

21. The 10 μF capacitor is open. Determine V_{dc} and $V'_{rms_{ac}}$.

22. The 10 μF and the 40 μF capacitors are open. Determine V_{dc} and $V'_{rms_{ac}}$.

23. All three capacitors are open. Determine V_{dc} and $V'_{rms_{ac}}$.

24. The 40 μF capacitor is open. The 20 H choke is shorted. Determine V_{dc} and $V'_{rms_{ac}}$.

25. The 20 H choke is shorted. The 80 μF capacitor is open. Determine V_{dc} and $V'_{rms_{ac}}$.

26. The 20 H choke is shorted. The 10 μF capacitor is open. Determine V_{dc} and $V'_{rms_{ac}}$.

27. All three capacitors are open. The 20 H choke is shorted. Determine V_{dc} and $V'_{rms_{ac}}$.

six

POWER
AMPLIFIERS

OPERATING CLASSIFICATION

The primary function of a power amplifier is to deliver power to a load as efficiently as possible. The load may be a speaker, antenna or any device that consumes power. Power amplifiers are used in public address (PA) systems, transmitters, receivers, record players and so forth.

Large input signals are used to provide appreciable amounts of output signal power. The dynamic characteristics of the BJT or FET types of semiconductors become nonlinear for such extremely wide variations in both current and voltage. Therefore, current conduction in the output circuit may not be continuous, with the result that small signal equivalent circuit analysis is valueless. Consequently, graphical techniques must be employed to evaluate the large signal operation of power amplifiers.

Power amplifiers can be operated as Class A, Class B or Class C. This classification system is based on the operating point location. The basic circuit of a Class A power amplifier is shown in Fig. 6-1a. The operating point is usually located midway between cutoff and saturation on the transistor characteristics, as shown in Fig. 6-1b.

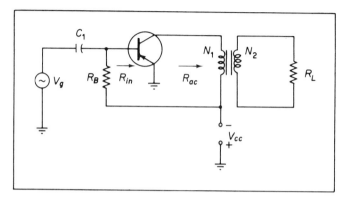

Figure 6-1a   *Class A Power Amplifier*

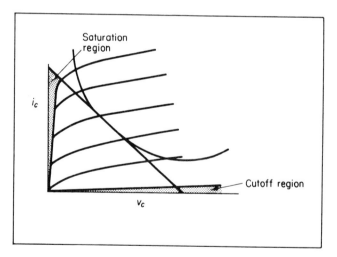

Figure 6-1b   *Operating Limits of Class A Power Amplifier*

In this circuit, $R_{in}C_1$ form the input coupling circuit. C_1 blocks any dc from the previous stage. The ac component is developed across R_B. In addition, R_B also acts as the biasing resistor used to maintain the transistor at the proper operating point. The output transformer acts as a collector load impedance and provides power transfer from the collector circuit to the speaker. The ac load impedance is equal to the turns ratio squared multiplied by the load resistor R_L. This ac load appears on the load line and defines the limits of operation of the transistor.

analysis A power amplifier is a power converter that changes dc power input to an ac power output. The input power is supplied by the source and is equal to $V_{cc}I_{cq}$. This power is utilized according to the following relationship:

$$V_{cc}I_{cq} = I_c^2 R_{ac} + \text{losses}$$

or

$$\text{dc power input} = \text{ac power output} + \text{dissipation across}$$
$$\text{the active device}$$

The ac load impedance into which the amplifying device feeds is defined by utilizing ideal transformer theory. The turns ratio N_1/N_2 is equal to a and is also defined by:

$$a = \frac{v_p}{v_s} = \frac{i_s}{i_p}$$

The impedance on the primary side is then given by:

$$R_{ac} = a^2 R$$

If there is no input signal, the output power is zero and the input power is dissipated across the amplifier device. Since designs are based on worst case conditions, the analysis of the system will assume the same.

The average output power dissipated across the primary impedance is:

$$P_{ac} = I_c^2 R_{ac}$$

The input power must be supplied by the power supply. Thus,

$$P_{in} = V_{cc}I_{cq}$$

The efficiency of conversion from dc power to ac power is defined by the relationship

$$\%\eta \ (\text{efficiency}) = \frac{\text{ac power output}}{\text{dc power input}} 100\%$$

For the assumed Class A case, the device characteristics are shown in Fig. 6-2.

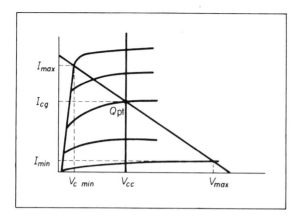

Figure 6-2   *Transistor Idealized Characteristics*

The output power can also be written as:

$$P_{ac} = \frac{(I_{max} - I_{min})^2}{(2\sqrt{2})^2} R_{ac}$$

The efficiency of power conversion can be written as:

$$\%\eta = \frac{(I_{max} - I_{min})^2}{8 V_{cc} I_{cq}} R_{ac} \; 100\%$$

Using a large signal swing for Class A amplifier operation, the minimum value of current is approximately zero. Thus,

$$I_{min} \cong 0$$

and

$$I_{cq} \cong \frac{I_{max}}{2}$$

Under these conditions, the power conversion efficiency becomes

$$\%\eta = \frac{I_{max} R_{ac}}{2 V_{cc}} 50\%$$

Again assuming maximum signal swing, $2V_{cc} \cong I_{max} R_{ac}$ so that the theoretical maximum value for the conversion efficiency is 50 percent. Actual operating efficiencies in a BJT amplifier approaches this maximum, usually operating from 38 to about 45 percent.

AMPLITUDE DISTORTION

A dynamic transfer curve which relates the input and output quantities can be used to determine the output waveform that results from a large input signal. Consider a large sinusoidal signal input. The resultant deviation of the output waveform from the sinusoidal waveform indicates non-linear distortion or the presence of undesirable harmonics. Refer to Fig. 6-3a. Note that two frequencies are shown whereby one is the fundamental and the other is the second harmonic. The sum of these two frequencies produces the resultant waveform shown in Fig. 6-3a(b).

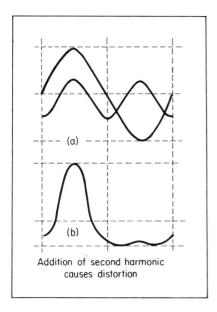

Addition of second harmonic
causes distortion

Figure 6-3a   *Addition of Second Harmonic Causes Distortion*

It is evident that a similar result is produced in a transistor circuit having the dynamic transfer characteristic shown in Fig. 6-3b. Note that additional harmonics are generated in the transistor circuit causing a distorted output. This action occurs because of the operation of the transistor in its nonlinear region. Note that the resultant output has a large positive output and a small negative half cycle.

An analytical method permits evaluation of the harmonic amplitudes by writing the equation for the transfer function of the device. The ac load line is used to establish the desired current and voltage values necessary to evaluate the amplitudes of the harmonic components.

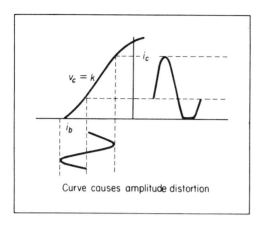

Curve causes amplitude distortion

Figure 6-3b *Amplitude Distortion*

The input signal to either device (the BJT or the MOSFET) may be considered as a cosine function. Then

BJT: $i_b - i_{b_q} = I_m \cos \omega t$

or

MOSFET: $e_g - e_{g_q} = E_m \cos \omega t$

Using trigonometric identities, some higher mathematics and some algebraic manipulation, the equation for the resultant output current becomes:

$$i_0 = I_{c_q} + A_0 + A_1 \cos \omega t + A_2 \cos 2\omega t + A_3 \cos 3\omega t + \cdots$$

where the A terms are the peak amplitudes of the harmonic components. The input signal can be assumed zero at some instant in time. The value of the transistor current at the operating point is defined as I_{c_q}. The subscript attached to the A coefficient defines the amplitude of the harmonic component. Thus,

Subscript	Harmonic
0	dc
1	fundamental
2	second
3	third
	etc.

Consider the load line that has been constructed on the BJT characteristics as shown in Fig. 6-4.

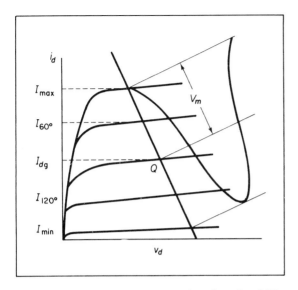

Figure 6-4 *Distortion Calculations from Load Line*

For most transistors, the terms above the fourth harmonic may usually be neglected. Consider as a first approximation and for simplicity of analysis a three-point schedule method. The output current equals:

$$i_c = I_{c_q} + A_0 + A_1 \cos \omega t + A_2 \cos 2\omega t$$

The three points selected from the characteristic curves are peak, average and minimum values. These values occur at:

$\omega t = 0$ yields maximum current

$\omega t = \dfrac{\pi}{2}$ yields average current

$\omega t = \pi$ yields minimum current

The corresponding values of current amplitudes or the harmonic coefficients can be obtained by:

ωt	$i_c = I_{c_q} + A_0 + A_1 \cos \omega t + A_2 \cos 2\omega t$
0	$I_{\max} = I_{c_q} + A_0 + A_1 + A_2$
$\dfrac{\pi}{2}$	$I_{c_q} = I_{c_q} + A_0 + 0 - A_2$
π	$I_{\min} = I_{c_q} + A_0 - A_1 + A_2$

It is evident that for the relationship $\omega t = 90°$, $A_0 = A_2$. The solution of the other two equations yields the following values.

$$A_1 = \frac{I_{\max} - I_{\min}}{2}$$

and

$$A_2 = \frac{I_{\max} + I_{\min} - 2I_{c_q}}{4}$$

It is possible to define percentage distortion due to a particular harmonic as the ratio of the harmonic amplitude to that of the fundamental multiplied by 100 percent. Thus,

$$\%D_2 = \frac{A_2}{A_1}100\% \qquad \%D_3 = \frac{A_3}{A_1}100\% \text{ etc.}$$

The average ac load power due to the fundamental only is given by:

$$P_{ac} = \tfrac{1}{2} A_1^2 R_{ac}$$

or

$$P_{ac} = \frac{(I_{\max} - I_{\min})^2}{8}R_{ac}$$

In terms of current and voltage, the resultant equation is:

$$P_{ac} = \frac{(V_{\max} - V_{\min})(I_{\max} - I_{\min})}{8}.$$

The additional dc component due to signal input will cause the ac load line to shift since the Q point has changed. New values of current can then be determined and the new "A" coefficients evaluated as a second order approximation. In general, the shift in the Q point is seldom large enough to disturb the calculations previously made.

The evaluation of the higher order harmonics utilizes a five-point schedule method. The approach is identical to that of the three-point schedule method, but requires additional terms.

The procedure for the five-point schedule method assumes the current equation is given by:

$$i_c = I_{c_q} + A_0 + A_1 \cos \omega t + A_2 \cos 2\omega t + A_3 \cos 3\omega t + A_4 \cos 4\omega t$$

Using the values for ωt as shown, the current equations then become:

$$\omega t = 0 \qquad\qquad I_{\max} = I_{c_q} + A_0 + A_1 + A_2 + A_3 + A_4$$

$$\omega t = 60° \qquad\quad I_{60°} = I_{c_q} + A_0 + \frac{1}{2}A_1 - \frac{A_2}{2} - A_3 - \frac{A_4}{2}$$

$$\omega t = \frac{\pi}{2} \qquad\qquad I_{c_q} = I_{c_q} + A_0 + 0 - A_2 + 0 + A_4$$

$$\omega t = 120° \qquad\;\; I_{120°} = I_{c_q} + A_0 - \frac{A_1}{2} - \frac{A_2}{2} + A_3 - \frac{A_4}{2}$$

$$\omega t = \pi \qquad\qquad I_{\min} = I_{c_q} + A_0 - A_1 + A_2 - A_3 + A_4$$

If we use algebraic techniques and elimination among these five equations, then, the resultant equations for the amplitudes of the harmonic components are:

$$A_0 = \frac{(I_{\max} + I_{\min}) + 2(I_{60°} + I_{120°})}{6} - I_{c_q}$$

$$A_1 = \frac{(I_{\max} - I_{\min}) + (I_{60°} - I_{120°})}{3}$$

$$A_2 = \frac{I_{\max} + I_{\min} - 2I_{c_q}}{4}$$

$$A_3 = \frac{(I_{\max} - I_{\min}) - 2(I_{60°} - I_{120°})}{6}$$

$$A_4 = \frac{(I_{\max} + I_{\min}) - 4(I_{60°} + I_{120°}) + 6I_{c_q}}{12}$$

The percentage third and fourth harmonic percentage distortion is given by:

$$\%D_3 = \frac{A_3}{A_1}100\%; \qquad \%D_4 = \frac{A_4}{A_1}100\%$$

The total harmonic distortion is defined as the ratio of the effective value of all the harmonics with respect to the fundamental. Thus,

$$D_T = \sqrt{D_2^2 + D_3^2 + D_4^2 + \cdots}$$

DESIGN CONSIDERATIONS

The application of large input signals to an amplifying device may introduce nonlinearities in the output and cause distortion. It is evident, therefore, that the output load selected, circuit bias and source impedance may yield the required output power but may have excessive distortion. The percentage distortion permissible in an amplifier is a function of the fidelity required. For audio frequency usage a low two percent distortion may be the maximum allowable distortion, whereas for radio receivers and public address systems, five percent distortion can be tolerated. The primary considerations of a power amplifying system are the power output and frequency requirements. These specifications govern the selection of the transistor. Consequently, once the transistor has been selected, additional manufacturer's data are known and the design procedure can then be applied.

STEPS IN THE DESIGN PROCEDURE

1. From manufacturer's data for $V_{c_{max}}$ and $P_{c_{rated}}$ construct a rated collector dissipation curve on the transistor characteristics shown in Fig. 6-5.

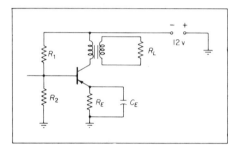

Figure 6-5 *Average Collector Characteristics of Types 2N301 and 2N301-A.*

2. For economical usage of the transistor, the maximum output occurs at a point tangent or slightly below tangency to the collector dissipation curve. Refer to Fig. 6-5. Construct ac load line as a test load. Its suitability will be determined by graphical analysis.

3. Choice of the dc operating point is arbitrary. Construct a vertical line from $V_{c_{max}}$ to the points of intersection with the ac load line and the collector dissipation curve and the dc load line has been constructed. Refer to Fig. 6-5.

4. From the ac load line, determine the values of I_{max}, I_{min}, V_{max}, V_{min}. These values are required to calculate the output power, input power and the percentage harmonic distortion.

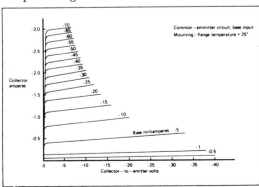

An illustrative problem will demonstrate the design procedure.

sample problem

A 2N301 transistor is used as a Class A transformer-coupled amplifier as shown in the figure below. The following parameters are given for the circuit:

$$V_{cc} = 12 \text{ V} \quad P_{c_{diss}} = 11 \text{ W} \quad R_L = 4 \text{ }\Omega$$

The characteristics for the 2N301 are shown in Fig. 6-6.

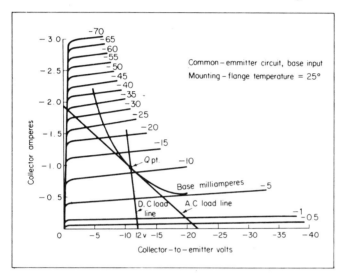

Figure 6-6   *Transformer-Coupled Amplifier.*

Solution:

Step 1:   On the transistor characteristics, construct a $P_{c_{diss}}$ curve.

Step 2:   Assume a drop across R_E to be about one volt. Consequently, the dc load line is constructed from $V_{cc} = 12$ V to a point of intersection of 11 V and the collector dissipation curve. This point yields the base bias current and the quiescent value of collector current. Thus,

$$I_{b_q} = 12 \text{ mA} \qquad I_{c_q} = 1.00 \text{ A}.$$

Step 3:   The base to emitter voltage is assumed to be 0.5 V. Since the emitter to ground voltage is 1 V dc, then the voltage from base to ground must equal 1.5 V. To determine the values of

R_1 and R_2 assume that the base bias resistance is 500 Ω. Then

$$\frac{V_{oc} - 1.5}{500} = 12 \text{ mA}$$

$$V_{oc} = 7.5 \text{ V}$$

$$V_{oc} = \frac{V_{cc}R_2}{R_1 + R_2} = \frac{V_{cc}R_T}{R_1}$$

Substituting values yields

$$R_1 = 1333 \ \Omega$$

The solution for R_2 is determined by the relationship:

$$R_2 = \frac{R_T R_1}{R_1 - R_T}$$

$$R_2 = 800 \ \Omega$$

The value of R_E is found from:

$$R_E = \frac{V_E}{I_{b_q} + I_{c_q}} = \frac{1}{1.00 + .012}$$

$$R_E = .99 \ \Omega$$

Step 4: An ac load line is constructed. The technique is to draw a line tangent to the collector dissipation curve and passing through the Q point. Assuming a maximum swing of 12 mA in the base circuit, the end points are:

$$V_{\max} = 21.5 \text{ V} \qquad I_{\max} = 1.55 \text{ A}$$
$$V_{\min} = 4.5 \text{ V} \qquad I_{\min} = 0$$

Step 5: Calculate P_i

$$P_i = V_{cc}I_{c_q}$$
$$P_i = 12 \ (1.00) = 12 \text{ watts}$$

Calculate P_o

$$P_o = \frac{(V_{\max} - V_{\min})(I_{\max} - I_{\min})}{8}$$

$$P_o = 3.29 \text{ W}$$

Calculate $P_{c_{diss}}$

$$P_{c_{diss}} = P_i - P_o$$
$$P_{c_{diss}} = 8.71 \text{ W}$$

Step 6:   Calculate R_{ac}
From graph, the ratio of voltage to current is:

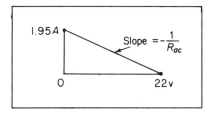

$$R_{ac} = \frac{21.5}{1.95} = 11.03 \ \Omega$$

Step 7:   Calculate the efficiency

$$\%\eta = \frac{P_o}{P_i} 100\% = \frac{3.29}{12} 100\%$$

$$\%\eta = 27.4\%$$

Step 8:   Calculate the turns ratio of the transformer

$$R_{ac} = a^2 R_L$$

$$R_{ac} = a^2 4 = 11.03$$

$$a = 1.66 \text{ turns ratio}$$

Step 9:   Calculate A_0, A_1 and A_2

$$A_0 = \frac{I_{max} + I_{min} - 2I_{c_q}}{4}$$

$$A_0 = .1125 \text{ A}$$

$$A_1 = \frac{I_{max} - I_{min}}{2}$$

$$A_1 = .775 \text{ A}$$

$$A_2 = A_0 = -.1125 \text{ A}$$

Step 10:   Calculate D_2

$$D_2 = \frac{A_2}{A_1} = 14.55\%$$

THERMAL CONSIDERATIONS

The maximum collector dissipation is a function of the transistor construction and thermal characteristics. For transistors in the milliwatt collector dissipation region, heat conduction along the leads or direct convection from the case to the surrounding air is adequate. For power transistors having outputs in the wattage region, heat removal is generally accomplished by mounting the transistor on a heat sink. Generally, the collector and shell are connected directly to the sink. In some cases a mica separator insulates the transistor from the sink, but permits heat dissipation by conduction.

A maximum allowable junction temperature is specified by the manufacturer for all power transistors. For germanium type transistors the maximum allowable junction temperature lies between 60° to 110° C. For silicon type transistors the maximum allowable junction temperature lies between 150° to 225° C.

The operating temperature of the junction is a function of the ambient temperature, the junction to ambient thermal resistance and the electrical power converted to heat plus stray thermal losses. The electrical analogue for the thermal equivalent circuit is shown in Fig. 6-7.

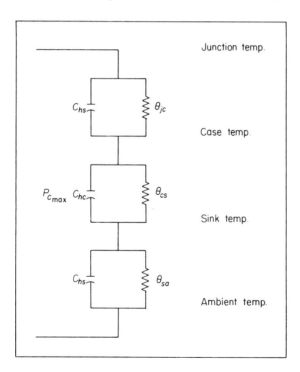

Figure 6-7   *Thermal-Electrical Analogy*

The maximum power dissipation capability can also be expressed in terms of the allowable junction temperature and is given by:

$$P_{c_{max}} = \frac{T_J - T_A}{\theta_{JA}} \text{ watts}$$

where:

T_J = junction temperature in °C.

T_A = ambient temperature in °C.

θ_{JA} = junction to ambient thermal resistance in °C/watt.

The overall thermal resistance θ_{JA} is evaluated by:

$$\theta_{JA} = \theta_{jc} + \theta_{cs} + \theta_{sa}$$

The thermal capacitance, C_h, is measured in watt-sec/°C. The product of the thermal capacitance and resistance yields a thermal time constant. The various thermal capacitances that exist within the transistor are useful as explanations for transient circuit phenomena. In worst case design, all capacitances are assumed zero.

Fig. 6-8 shows a plot of the maximum power dissipation capability of an RCA 2N301 power transistor as a function of temperature for various values of circuit thermal resistances.

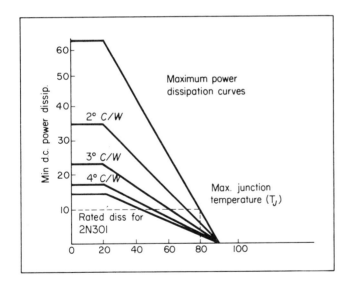

Figure 6-8   *Power vs Temperature Derating Curve*

Note that substantial amounts of power can be dissipated in a power transistor when the ambient temperature and thermal resistance are both maintained low. Most manufacturers specify a maximum collector dissipation at high temperatures. Thus, the 2N301 transistor has a maximum collector dissipation of 15 watts at an ambient temperature of 80° C.

A typical transformer coupled Class A amplifier using a 2N2869/2N301 transistor is shown in Fig. 6-9.

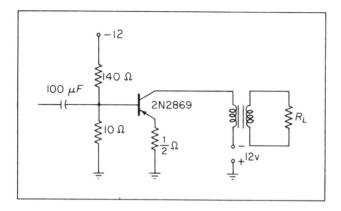

Figure 6-9 *A Single Ended Class A Power Amplifier*

VOLTAGE LIMITATIONS ON THE TRANSISTOR

Power output from an amplifier increases with applied voltage. There are two physical limitations on transistors.

The first is called *punchthrough*. As voltage is increased across the collector to base junction, the depletion region becomes wider, narrowing the base region. At some high collector voltage, the collector depletion region may extend completely through the base and short across the emitter.

Punchthrough of the base does not damage the transistor if the current is limited, but the transistor becomes inoperative above the limiting voltage.

First breakdown occurs in a transistor with the avalanching of collector current at some collector voltage. This does not seriously damage the transistor. However, *second breakdown*, which can follow avalanching, can lead to transistor destruction as shown in Fig. 6-10.

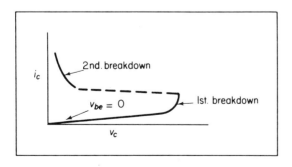

Figure 6-10 *Volt Ampere Curve Illustrating Breakdowns*

Safe operating voltage and current limits are usually specified by charts for each power transistor type.

CLASS A PUSH-PULL AMPLIFIER

A push-pull power amplifier is formed by combining two single ended amplifiers as shown in Fig. 6-11.

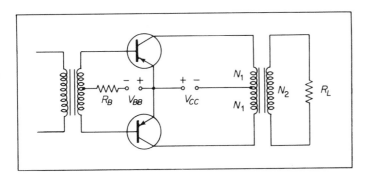

Figure 6-11 *Push-Pull Power Amplifier*

The push-pull power amplifier produces more than double the power output of a single ended amplifier and does it with considerably less distortion. It possesses the advantage of having dc collector current flowing through the transformer in opposite directions eliminating the problem of transformer core saturation.

theory The signal input is introduced through a center tapped transformer. Thus, when the signal voltage to the base of Q_1 is positive, the signal voltage at the base of Q_2 is negative. The bias current is increased in Q_2 and reduced in Q_1. Reversal of polarity of signal swing increases the bias current in Q_1 with a consequent reduction in bias current of Q_2. This push-pull operation is reflected in the output circuit as increased collector current swing.

analysis The characteristics of a push-pull circuit can be demonstrated by the following analysis. Assuming that both transistors have identical characteristics, the output current of either transistor is given by:

$$i_o = I_{cq} + a_1 v_s + a_2 v_s^2 + a_3 v_s^3 + \cdots$$

It is evident from the circuit that the input voltages are:

$$v_{t_1} = V_m \sin \omega t$$

and

$$v_{t_2} = V_m \sin (\omega t + \pi)$$

The output current of Q_1 is then:

$$i_{o_1} = I_{cq_1} + a_1 V_m \sin \omega t + a_2 V_m^2 \sin^2 \omega t + a_3 V_m^3 \sin^3 \omega t + \cdots$$

and the output current of Q_2 is given by:

$$i_{o_2} = I_{cq_2} + a_1 V_m \sin (\omega t + \pi) + a_2 V_m^2 \sin^2 (\omega t + \pi)$$
$$+ a_3 V_m^3 \sin_3 (\omega t + \pi) \cdots$$

By the use of trigonometric identities and some algebraic manipulation, the currents can be simplified to:

$$i_{o_1} = I_{q_1} + B_0 + B_1 \sin \omega t - B_2 \cos 2\omega t + B_3 \sin 3\omega t - B_4 \cos 4\omega t$$

and

$$i_{o_2} = I_{q_2} + B_0 + B_1 \sin (\omega t + \pi) - B_2 \cos 2(\omega t + \pi)$$
$$+ B_3 \sin 3(\omega t + \pi) - B_4 \cos 4(\omega t + \pi) \cdots$$

Since: $\sin (\omega t + \pi) = -\sin \omega t$

$\cos 2(\omega t + \pi) = \cos 2\omega t$

$\sin 3(\omega t + \pi) = -\sin 3\omega t$

$\cos 4(\omega t + \pi) = \cos 4\omega t$

Using these identities, current i_{o_2} becomes

$$i_{o_2} = I_{q_2} + B_0 - B_1 \sin \omega t - B_2 \cos 2\omega t - B_3 \sin 3\omega t - B_4 \cos 4\omega t$$

The total current in the output is equal to the difference of the two currents. Thus,

$$i_T = i_{o_1} - i_{o_2}$$

and

$$i_T = B_1 \sin \omega t + B_3 \sin 3\omega t + B_5 \sin 5\omega t + \cdots$$

Note that all even harmonics are eliminated in the output. In addition, any power supply ripple that exists will also be eliminated. Fig. 6-12 illustrates the composite output characteristics and the graphical results of Class A push-pull operation.

A typical push-pull Class A power amplifier is shown in Fig. 6-13. It is evident that collector current flows continuously and the transistor dissipation is greatest when no ac signal is present.

The power transistors also require heat sinks for proper operation. Circuits of this type are seldom used because transistors function much more efficiently in Class B operation.

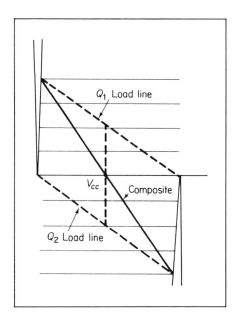

Figure 6-12 *Composite Output Characteristics*

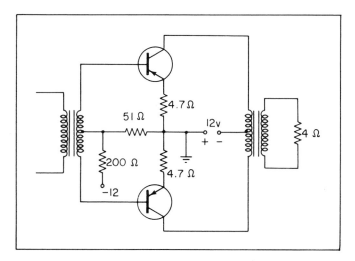

Figure 6-13 *Push-Pull Class A Power Amplifier*

CLASS B PUSH-PULL AMPLIFIER

For an increased power efficiency and the greater output possible for the same input power, Class B operation is extremely desirable. Fig. 6-14 shows a typical Class B push-pull power amplifier stage.

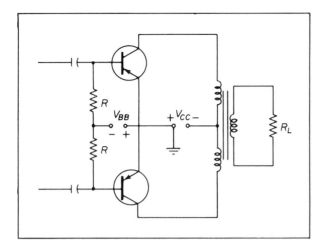

Figure 6-14 *Class B Push-Pull Power Amplifier*

In Class B operation, the transistors are biased almost to cutoff so that at zero signal input, the collector current and collector dissipation both are reduced to zero. A dynamic transfer characteristic is plotted in Fig. 6-15. The transfer characteristics are placed back to back for the zero base to

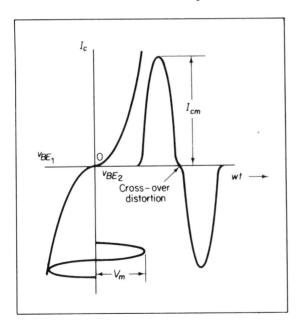

Figure 6-15 *Dynamic Transfer Characteristic Showing Crossover Distortion*

emitter bias conditions. This is evident in Fig. 6-15. Note that application of a large base to emitter signal input results in the dynamic collector current swing shown in Fig. 6-15. It is evident that the collector currents produced do not match properly, with consequent distorted output called "crossover distortion."

To remove the improper joining of the current waveforms, a small forward base to emitter dc bias voltage is applied to the transistor. A convenient method for determining the required bias voltage is to project linearly from the transfer curve to cutoff as shown in Fig. 6-16. The use of this projected cutoff bias voltage eliminates *crossover distortion*. Typical value of push-pull bias is about 0.2 V.

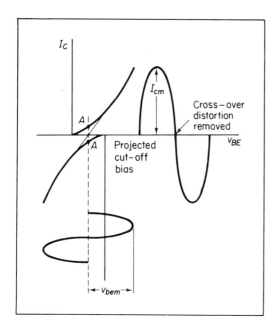

Figure 6-16   *Effect of Forward Bias on Crossover Distortion*

analysis   The analysis of a Class B amplifier can be performed by first assuming ideal conditions. Thus, the dc input current to the collector is given by:

$$I_{c_q} = \frac{2I_m}{\pi}$$

The input dc power to the collector is:

$$P_{dc} = \frac{2V_{cc}}{\pi} I_m$$

The output current is given by:

$$I_{rms} = \frac{I_m}{\sqrt{2}}$$

The output power is evaluated by the formula:

$$P_{ac} = \frac{I_m^2 R_{ac}}{2}$$

If it is assumed that the output voltage is approximately equal to the supply voltage, then

$$R_{ac} = \frac{V_{cc}}{I_m}$$

and $$P_{ac} = \frac{V_{cc}}{2} I_m$$

The maximum collector efficiency using these ideal relationships is given by:

$$\%\eta_{max} = \frac{P_{ac}}{P_{dc}} 100\% = \frac{\pi}{4} 100\%$$

$$\%\eta_{max} = 78.5\%$$

That Class B operation is considerably better than Class A is shown by the consequent widespread usage of the Class B amplifier.

Linearity of output for large input signals is dependent on the linearity of the transistor characteristics. In Class B push-pull operation, one transistor does not compensate for the other as is done in Class A operation. For example, saturation effects from large signal input can occur with resultant distortion.

An illustrative problem will demonstrate the theory.

sample problem

A Class B push-pull power amplifier using a transistor having the given characteristic (refer to Fig. 6-17) has:

$$V_{cc} = 14 \text{ V} \quad P_{c_{diss}} = 30 \text{ W}$$

Find: (a) P_o (b) P_i (c) η

Solution:

Step 1:   On the characteristics, construct a $P_{c_{diss}}$ curve equal to 30 W. From V_{cc} equal to 14 V, construct a line tangent to the

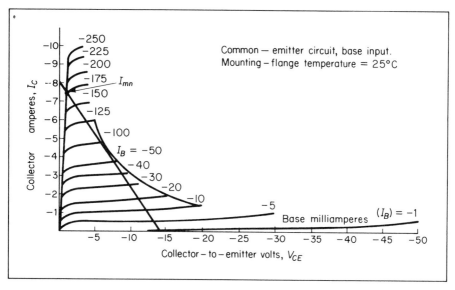

Figure 6-17   *Typical Collector Characteristics for Types 2N2869/2N301 and 2N2870/2N301A*

collector dissipation curve. The load line has been constructed. The value of R_{ac} is given by:

$$R_{ac} = \frac{14 - 1}{7.6 - 0} = 1.71 \; \Omega$$

Step 2:   Calculate the output power

$$P_o = \frac{13 \times 7.6}{2}$$

$$P_o = 49.4 \; W$$

Step 3:   Calculate the input power

$$P_i = 14 \times 2 \times \frac{7.6}{\pi}$$

$$P_i = 67.7 \; W$$

Step 4:   Calculate the collector circuit efficiency

$$\%\eta = \frac{49.4}{67.7} 100\%$$

$$\%\eta = 73\%$$

A typical transistor Class B amplifier is shown in Fig. 6-18.

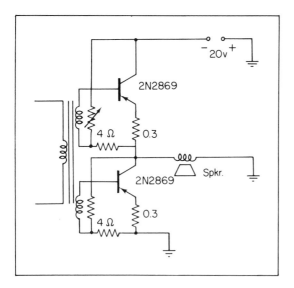

Figure 6-18   *A Typical 15 W Class B Amplifier*

PHASE INVERTERS

Two voltages equal in magnitude but 180° out of phase are needed for push-pull amplifier input. Circuits that provide the necessary voltages and phase relationships are called *phase inverters*. The simplest form of a phase inverter is shown in Fig. 6-19.

The signals at points A and B with respect to ground are 180° out of phase, but generally signal A is greater than signal B. To provide equal

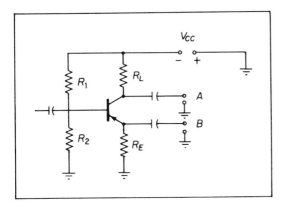

Figure 6-19   *Phase Inverter*

magnitudes and to maintain the phase relationships, an additional transistor is added as shown in Fig. 6-20.

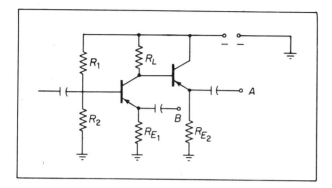

Figure 6-20 *Two-Stage Phase Inverter*

TROUBLESHOOTING ANALYSIS

The most common cause of circuit failure in a push-pull power amplifier is the failure of one of the amplifying devices. It is highly improbable that both amplifying devices of the push-pull amplifier should fail simultaneously. Consequently, a no output condition means:

(a) Failure of a compoment element that is common to both amplifying devices. Thus the emitter resistor or output transformer can be open.

(b) Power supply circuit open.

(c) No input signal. This indicates a circuit failure prior to the output stage.

If one of the two amplifying devices fails, a reduced output will result. The amplifiers can be improperly biased because of changes in component values resulting in either a reduced or distorted output. An oscilloscope should be used to determine the source of distortion or mismatch condition.

CLASS C AMPLIFIER

A Class C amplifier is an amplifier biased from 1.5 to 4 times cutoff bias. The collector current is zero when no signal is applied. The application of base input signal causes collector current to flow for appreciably less than 180°. Class C amplifiers are characterized by high collector circuit efficiency and are used to develop rf output powers when a linear relationship is not a system requirement between input and output signals. A typical Class C amplifier is shown in Fig. 6-21.

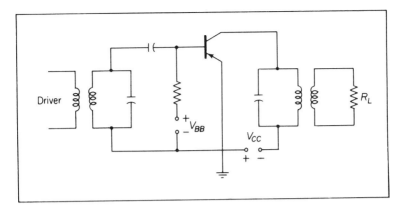

Figure 6-21 *Class C Amplifier*

The transistor acts as a switch and the collector supply delivers energy to the amplifier only when the largest portion of the energy will be transferred to the load.

The tuned circuit or *tank circuit* maintains energy transfer between the coil and the capacitor in an oscillatory manner. If a pulse of energy is fed to the system once each cycle to replace the energy lost over one cycle, then the "flywheel effect" of the tank circuit carries the system through one complete cycle until the next input pulse.

problems

1. A Class A transformer-coupled power amplifier uses a $30:1$ transformer to drive a 4 Ω load. Find the ac primary impedance.

2. Determine the turns ratio of a transformer used to couple a primary impedance of 6 kΩ to an 8 Ω load.

3. A transformer-coupled Class A power amplifier drives an 8 Ω speaker through a $20:1$ transformer. The dc supply voltage is 24 V and the circuit delivers 4 W to the load. Calculate:
   (a) primary output power.
   (b) *rms* value of primary voltage and current.
   (c) *rms* value of secondary voltage and current.

4. Calculate the efficiency of the circuit of problem 3 if the operating collector current is 300 mA.

5. The following data are given:
$$I_{max} = 2.7 \text{ A} \qquad I_{c_q} = 2.0 \text{ A}$$
$$I_{min} = 1.4 \text{ A}$$
Calculate the percentage second harmonic distortion.

6. The following data are given:
$$I_{max} = 4.8 \text{ A} \qquad I_{60°} = 3.6 \text{ A}$$
$$I_{min} = 1.0 \text{ A} \qquad I_{120°} = 1.5 \text{ A}$$
$$I_{c_q} = 3.0 \text{ A}$$
Determine: (a) $\%D_2$ (b) $\%D_3$ (c) $\%D_4$ (d) D_T

7. A transformer-coupled Class A power amplifier supplies 1.2 W to a 4 kΩ load. The zero signal dc collector current is 40 mA and the dc current with signal is 50 mA. Determine the percentage second harmonic distortion.

8. A transformer-coupled Class A power amplifier supplies 6 W to a speaker load of 4 Ω. The zero signal collector current is 1.5 A. The dc collector current with signal is 1.7 A. Determine the percentage second harmonic distortion.

9. The following data are given.
$$I_{max} = 600 \text{ mA} \qquad I_{60°} = 400 \text{ mA}$$
$$I_{min} = 25 \text{ mA} \qquad I_{120°} = 75 \text{ mA}$$
$$I_{c_q} = 250 \text{ mA}$$
Find: (a) percent second harmonic distortion
   (b) percent third harmonic distortion
   (c) percent fourth harmonic distortion
   (d) total distortion.

10. A 2N301 transistor having the characteristics shown in Fig A-1 (in Appendix) is used as a Class A power amplifier. The following data are specified.

$$V_{cc} = -15 \text{ V} \qquad P_{c_{diss}} = 15 \text{ W} \qquad V_E = 1.5 \text{ V}$$
$$R_L = 4 \text{ }\Omega \qquad V_{be} = 1 \text{ V} \qquad R_{th} = 500 \text{ }\Omega$$

Find: (a) R_{ac} (b) R_E (c) R_1 and R_2 (d) P_L (e) $\%D_2$ (f) $\%\eta$.

seven

FEEDBACK AMPLIFIERS

INTRODUCTION

For a given amplifier system, the circuit properties such as amplification, frequency characteristics and input and output impedances are fixed. In some circuit applications it may be desirable to alter some of the properties of the given amplifier; e.g., input and output impedances and so forth. This can be done by the application of feedback techniques.

Feedback is said to exist in a circuit when a portion of the output is returned or *fed back* to the input and combined with the input signal. The new amplifier circuit formed will have properties uniquely different from those of the original amplifier.

FEEDBACK

In its simplest form, the effects of feedback are readily evident by considering the block diagram shown in Fig. 7-1. Note that the output of the feedback network is placed in series with the input signal. Various techniques exist for network coupling, but the series technique is used for simplicity.

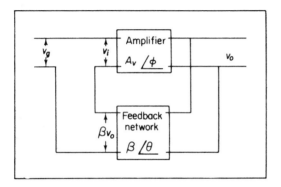

Figure 7-1   *Block Diagram of a Feedback Amplifier*

theory
1. The signal is applied to the input terminals of the amplifier.
2. The output signal voltage is v_0.
3. This output voltage is applied to the input terminals of the feedback network.
4. The output voltage of feedback network, known as the feedback voltage, is returned to the input. This process is repetitive.

analysis   The relationship for the input voltage to the amplifier is given by:

$$v_i = v_g + \beta v_0$$

The voltage amplification of the amplifier is given by:

$$A_v = \frac{v_0}{e_i}$$

The system amplification is given by (symbolized by A_{v_r}):

$$A_{v_r} = \frac{v_0}{v_g}$$

Thus, combining equations as follows:

$$v_i = \frac{v_o}{A_v} \quad \text{and} \quad v_g = v_i - \beta v_o$$

The resultant equation is:

$$A_{v_r} = \frac{v_o}{\dfrac{v_o}{A_v} - \beta v_o}$$

and

$$A_{v_r} = \frac{A_v}{1 - A_v \beta}$$

This particular equation expresses the feedback voltage amplification in terms of A_v, the voltage amplification (called the open loop voltage amplification) without feedback and a feedback network factor β. The amount of feedback is determined by β or the ratio of the feedback voltage to the output voltage. Thus,

$$\beta = \frac{v_{fb}}{v_o}$$

It is evident that since A_v and β are both complex numbers, the resultant voltage amplification will also be complex in nature.

The feedback is termed negative or degenerative when it reduces the magnitude of the voltage amplification. It is termed positive or regenerative when it increases the magnitude of the voltage amplification. It is obvious that when the value of $A_v \beta$ is exactly equal to unity at a phase angle of zero, that the system amplification has a zero in the denominator resulting in uncontrolled oscillation.

Consequently, the following properties may be ascribed to negative feedback.

1. The amplifier frequency response is usually extended.

2. The system voltage amplification is reduced.

3. The harmonic distortion existent within the output signal is usually reduced.

The analysis of the frequency response of the typical RC coupled amplifier with negative feedback is performed in the following manner.

The low frequency voltage amplification of the typical RC coupled amplifier is given by:

$$A_{v_{10}} = \frac{A_{v_{mid}}}{1 - j\frac{f_1}{f}}$$

The low frequency voltage amplification of the same system, but with feedback, is given by:

$$A_{v_{10r}} = \frac{A_{v_{10}}}{1 - A_{v_{10}}\beta}$$

Substituting and simplifying the feedback equation becomes:

$$A_{v_{10r}} = \frac{A_{vr_{mid}}}{1 - j\frac{f_1'}{f}}$$

where:

$$A_{v_{mid_r}} = \frac{A_{v_{mid}}}{1 - A_{v_{mid}}\beta}$$

$$f_1' = \frac{f_1}{1 - A_{v_{mid}}\beta}$$

Although $A_{v_{mid}}$ and β are both complex numbers, the value of the half power point frequency at the low end is a real value only. The half power point frequency at the high end of the amplifier frequency response can be determined in a similar manner. Thus:

$$A_{v_{hi}} = \frac{A_{v_{mid}}}{1 + j\frac{f}{f_2}}$$

The system amplification with feedback is given by:

$$A_{v_{rhi}} = \frac{A_{v_{hi}}}{1 - A_{v_{hi}}\beta}$$

The resultant equation after substituting and simplifying is:

$$A_{v_{rhi}} = \frac{A_{v_{rmid}}}{1 + j\frac{f}{f_2'}}$$

where:

$$f_2' = f_2(1 - A_{v_{mid}}\beta)$$

An illustrative problem will demonstrate the theory.

sample problem

An RC coupled amplifier has a midfrequency voltage amplification of −20 and a frequency response from 100 Hz to 20 kHz. A feedback network with $\beta = 0.2$ is incorporated into the circuit. Determine the new system performance.

Solution:

Step 1: Calculate A_{v_r}

$$A_{v_r} = \frac{-20}{1 - (-20).2}$$

$$A_{v_r} = -4$$

Step 2: Calculate f_1' and f_2'

$$f_1' = \frac{100}{1 - (-20).2}$$

$$f_1' = 20 \text{ Hz}$$

$$f_2' = f_2[1 - (-20).2]$$

$$f_2' = 20(5) \text{ kHz}$$

$$f_2' = 100 \text{ kHz.}$$

Step 3: A comparison of the old and new system performances is given below.

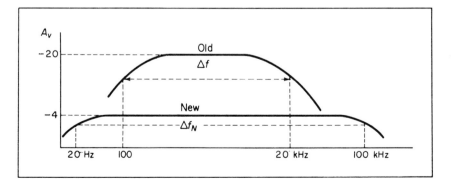

SERIES FEEDBACK

A common emitter amplifier that uses series feedback is shown in Fig. 7-2. In this circuit, current feedback is obtained by the insertion of an unbypassed resistor R_E. Resistor R_E is small compared to the load resistor R_L, and, consequently, has little effect on the load current.

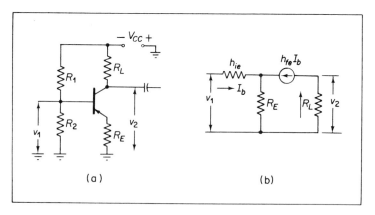

(a)   (b)

Figure 7-2   (*a*) *Series Feedback* (*b*) *Equivalent Circuit*

analysis   The input and output circuit equations are:

$$v_1 = I_b h_{ie} + (I_b + I_c)R_E$$

or $\quad v_1 = [h_{ie} + (1 + h_{fe})]R_E I_b$

and $\quad v_2 = -I_c R_L = -h_{fe} I_b R_L$

The voltage amplification of the circuit is:

$$A_v = \frac{v_2}{v_1}$$

$$A_v = \frac{-h_{fe}R_L}{h_{ie} + (1 + h_{fe})\,R_E}$$

The definition of the feedback factor β, then, is given by:

$$\beta \cong \frac{(1 + h_{fe})R_E}{h_{fe}R_L} \cong \frac{R_E}{R_L}$$

provided that:

(a) $(1 + h_{fe})R_E \gg h_{ie}$

(b) $h_{fe} \gg 1$

The addition of the term $(1 + h_{be})R_E$ to h_{ie} reduces the overall system amplification. The input resistance with series feedback is given by:

$$\frac{v_1}{I_b} = R_{in} = h_{ie} + (1 + h_{fe})R_E$$

Note that the input resistance has increased while the voltage amplification and system distortion have decreased. Series feedback is most effective when the driving source impedance is low. Otherwise, a signal input with a high source resistance tends to be a constant current source and the feedback system has slight effect on the circuit operation.

SERIES FEEDBACK (FET)

The variations in amplification caused by the differences in FETs of same type number are so great that feedback must be used to stabilize the amplification of the device. The basic circuit of a series feedback FET amplifier is shown in Fig. 7-3.

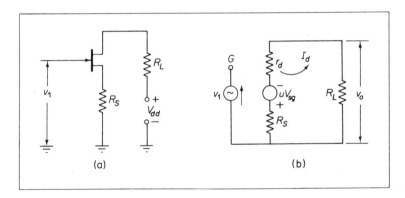

Figure 7-3   (*a*) *Series Feedback* (*b*) *Equivalent Circuit*

analysis   The source to gate voltage of the FET is given by:

$$V_{sg} = V_1 - I_d R_s$$

The drain current is equal to:

$$I_d = \frac{+uV_1}{r_d + R_L + R_s(1 + u)}$$

The output voltage is given by:

$$V_0 = \frac{-uV_1 R_L}{r_d + R_L + R_s(1 + u)}$$

The voltage amplification of the system is:

$$A_{v_r} = \frac{-uR_L}{r_d + R_L + (1 + u)R_s}$$

The feedback technique can also be used and is applied as follows. The first step is to determine the feedback factor β:

$$\beta = \frac{I_d R_s}{I_d R_L}$$

The open loop voltage or nonfeedback amplification is given by:

$$A_v = \frac{-uR_L}{r_d + R_L + R_s}$$

Substituting these values into the basic feedback equation yields:

$$A_{v_r} = \frac{A_v}{1 - A_v\beta}$$

and $$A_{v_r} = \frac{\dfrac{-uR_L}{r_d + R_L + R_s}}{1 - \left(\dfrac{R_s}{R_L}\right)\left(\dfrac{-uR_L}{r_d + R_L + R_s}\right)}$$

Simplifying, yields the resultant solution:

$$A_{v_r} = \frac{-uR_L}{r_d + R_L + (1 + u)R_s}$$

The two results are identical.

It is also evident that the drain resistance of the FET has increased tremendously. The new drain resistance is given by:

$$r_{d_r} = r_d + R_s(1 + u)$$

An illustrative problem will demonstrate the theory.

sample problem

The following circuit is given.

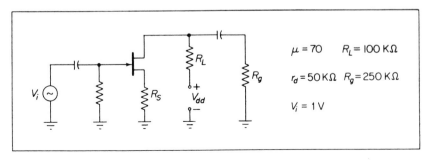

$u = 70$ $R_L = 100 \text{ k}\Omega$
$r_d = 50 \text{ k}\Omega$ $R_g = 250 \text{ k}\Omega$
$V_i = 1 \text{ V}$

Find: (a) R_s for $A_{v_r} = -20$
        (b) A_{v_r} if $R_s = 1 \text{ k}\Omega$
        (c) r_{d_r} if $R_s = 1 \text{ k}\Omega$

Solution:

   Step 1:   Calculate R_{ac}

$$R_{ac} = \frac{R_L R_g}{R_L + R_g} = 71.4 \text{ k}\Omega$$

   Step 2:   Solve for R_s

$$A_{v_r} = \frac{-u R_{ac}}{r_d + R_{ac} + R_s(1 + u)}$$

$$-20 = \frac{-70(71.4)}{50 + 71.4 + 71 R_s}$$

$$R_s = 1.815 \text{ k}\Omega$$

   Step 3:   Solve for A_{v_r} with $R_s = 1 \text{ k}\Omega$

$$A_{v_r} = \frac{-u R_{ac}}{r_d + R_{ac} + R_s(1 + u)}$$

$$A_{v_r} = \frac{-70(71.4)}{50 + 71.4 + 71}$$

$$A_{v_r} = -26$$

   Step 4:   Solve for r_{d_r}

$$r_{d_r} = r_d + R_s(1 + u)$$

$$r_{d_r} = (50 + 71)10^3 \ \Omega$$

$$r_{d_r} = 121 \text{ k}\Omega$$

VOLTAGE FEEDBACK

A shunt or voltage feedback circuit is illustrated in Fig. 7-4.

analysis   In the analysis of shunt feedback circuits, the following assumptions will be made.

1. R_1 in parallel with R_2 is extremely large and is considered to be open

2. R_F is large compared to R_L. Thus, $I_L R_L \cong I_c R_L$

3. $R_F \gg h_{ie}$

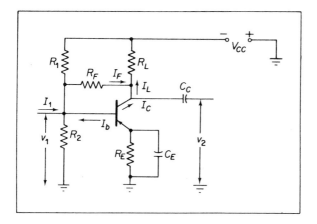

Figure 7-4   *Shunt Feedback*

Thus, from the equivalent circuit shown in Fig. 7-5, the following equations may be written:

$$I_F = \frac{R_L}{h_{ie} + R_F} I_c$$

or $$I_F \cong \frac{R_L}{R_F} I_c$$

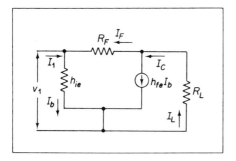

Figure 7-5   *Equivalent Circuit*

The input current I_1 is given by:

$$I_1 = I_b - I_F$$

Note that, since $I_c = h_{ef} I_b$, then

$\dfrac{I_c}{I_1} = A_{i_r}$. The value of A_{i_r} is also equal to:

$$A_{i_r} = \frac{A_{i_{mid}}}{1 - A_{i_{mid}}\left(\dfrac{R_L}{R_F}\right)}$$

It is evident that the feedback factor, β, is given by the relationship:

$$\beta = \frac{R_L}{R_F}$$

The input resistance is given by the ratio of the input voltage to the input current. Thus,

$$R_{in_r} = \frac{v_i}{I_1}$$

also $I_1 = I_b - \left(\dfrac{R_L}{R_F}\right)I_c$

and $I_1 = I_b(1 - A_i\beta)$

$$R_{in_r} = \frac{R_{in}}{1 - A_{i_{mid}}\beta}$$

where:

R_{in} = input resistance without feedback.

It is evident that although the frequency response has increased with reduction in amplification, the input resistance is decreased in voltage or shunt feedback.

SHUNT FEEDBACK (FET)

A circuit using shunt or voltage feedback that incorporates an FET is shown in Fig. 7-6.

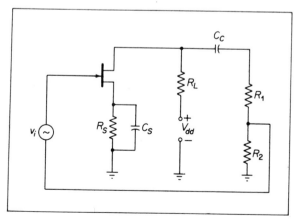

Figure 7-6 *FET Voltage Feedback Circuit*

The simplest method of circuit analysis is to apply the feedback technique. Thus, for the equivalent circuit shown in Fig. 7-7, the feedback factor, β, is given by:

$$\beta = \frac{R_2}{R_1 + R_2}$$

and $$R_{ac} = \frac{R_L(R_1 + R_2)}{R_L + R_1 + R_2}$$

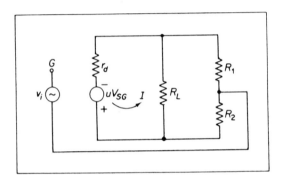

Figure 7-7   *Equivalent Circuit*

The open loop voltage amplification of the circuit is given by:

$$A_v = \frac{-uR_{ac}}{r_d + R_{ac}}$$

Substituting the value of β and A_v into the feedback equation yields:

$$A_{v_r} = \frac{\dfrac{-uR_{ac}}{r_d + R_{ac}}}{1 - \beta\left(\dfrac{-uR_{ac}}{r_d + R_{ac}}\right)}$$

and $$A_{v_r} = \frac{-uR_{ac}}{r_d + R_{ac}(1 + u\beta)}$$

If both numerator and denominator are divided by the term $(1 + u\beta)$ the resultant equation becomes:

$$A_{v_r} = \frac{-\left(\dfrac{u}{1 + u\beta}\right)R_{ac}}{\left(\dfrac{r_d}{1 + u\beta}\right) + R_{ac}}$$

or $\quad A_{v_r} = \dfrac{-u_r R_{ac}}{r_{d_r} + R_{ac}}$

Note that both the amplification factor and the FET drain resistance are decreased in voltage feedback.

COMPOUND FEEDBACK (FET)

When both current and voltage feedback are combined in an amplifier, the circuit is called compound feedback. A circuit of this type is shown in Fig. 7-8.

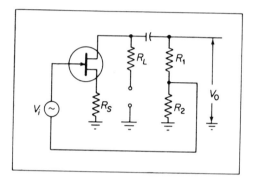

Figure 7-8 *Compound Feedback Circuit*

analysis Using the feedback technique, the overall feedback factor, β, is given by:

$$\beta = \beta_1 + \beta_2$$

where:

$$\beta_1 = \frac{R_s}{R_{ac}} \quad \text{and} \quad \beta_2 = \frac{R_2}{R_1 + R_2}$$

The open loop voltage amplification is given by:

$$A_{v_r} = \frac{-u R_{ac}}{r_d + R_s + R_{ac}}$$

The resultant voltage amplification is given by:

$$A_{v_r} = \frac{\dfrac{-u R_{ac}}{r_d + R_s + R_{ac}}}{1 - (\beta_1 + \beta_2)\left(\dfrac{-u R_{ac}}{r_d + R_s + R_{ac}}\right)}$$

Simplification of the feedback equation yields:

$$A_{v_r} = \frac{-uR_{ac}}{r_d + R_s(1 + u) + R_{ac}(1 + u\beta_2)}$$

Dividing the numerator and denominator by the term $(1 + u\beta_2)$ results in the following equation:

$$A_{v_r} = \frac{-u'_r R_{ac}}{r'_{d_r} + R_{ac}}$$

where:

$$u'_r = \frac{u}{1 + u\beta_2}$$

$$r'_{d_r} = \frac{r_d + R_s(1 + u)}{1 + u\beta_2}$$

A typical FET current feedback amplifier is shown in Fig. 7-9.

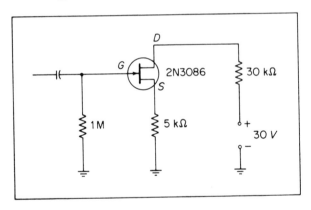

Figure 7-9 *FET Feedback Amplifier*

OPERATIONAL AMPLIFIER

An amplifier capable of performing the basic mathematical operations of addition, subtraction, differentiation and integration is called the operational amplifier. This performance is achieved by a feedback amplifier as shown in Fig. 7-10.

The amplifier has an input element, Z_i, and a feedback element, Z_F. The feedback element may be either an R, L, C or various combinations thereof.

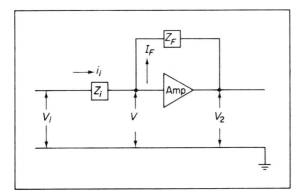

Figure 7-10 *Operational Amplifier*

If the current into the amplifier is made negligible, then $i_i = i_F$ and

$$\frac{v_i - v_s}{Z_i} = \frac{v_s - v_2}{Z_F}$$

Assuming a 180° phase shift through the amplifier, the voltage amplification is:

$$A_v = \frac{-v_2}{v_s}$$

The output voltage, v_2, is equal to:

$$v_2 = \left(\frac{-Z_F}{Z_i}\right)\left(\frac{v_i}{1 + \dfrac{Z_F}{Z_i}\dfrac{1}{A_v}}\right)$$

If the voltage amplification is made extremely large compared to unity, then

$$1 \gg \frac{Z_F}{Z_i A_v}$$

and $\quad v_2 = -\left(\dfrac{Z_F}{Z_i}\right) v_i$

This equation shows that the input and output voltages are related as the negative of Z_F/Z_i. If, for example, $Z_F = R_F$ and $Z_i = R_i$, then

$$v_2 = -\frac{R_F}{R_i} v_i$$

The two possibilities are:

1. Let $R_F = R_i$, then the circuit acts as a sign changer or phase inverter.

2. Let $R_F = n R_i$, then the system amplification has a voltage amplification of "n." If n is less than unity, the circuit becomes an attenuator, since this can be accomplished by circuit resistors rather than amplifiers. Consequently, this possibility was not considered.

Another useful operational circuit uses $Z_F = \dfrac{1}{j \omega C_F}$ and $Z_i = R$, then the value of v_2 is:

$$v_2 = \frac{-1}{j \omega R C_F} v_i$$

The output voltage is equal to the negative of the integral of the input multiplied by a constant equal to $\dfrac{1}{R C_F}$.

TONE CONTROL CIRCUITS

In many low-level amplifier applications some type of frequency compensating networks are used to boost either the low frequency or high frequency or both simultaneously. The frequency range of a phonograph record or magnetic tape depends on various factors such as composition, electrical and mechanical characteristics of the recording equipment, speed of rotation and so forth.

To achieve a wide frequency range of operation the manufacturers of commerical recordings utilize equipment that produces a variable amplitude with respect to frequency. To insure proper and uniform reproduction of a high fidelity recording, the electronic portion of the recording system must provide proper compensation for the low and high frequency variations.

One of the functions of a feedback circuit is to shape the response characteristic to meet specific circuit requirements. Consider the circuit that provides a *bass boost*, shown in Fig. 7-11.

The feedback loop from the collector to the base consists of a series RC network. Since capacitor, C, has a different reactance for different frequencies, the voltage feedback is variable. By selecting the value of C so that it will have a high reactance to the low or bass frequencies, but a relatively low reactance to the middle and high frequencies, the feedback circuit can be made to return very little signal at the low frequencies so that the lower the frequency applied, the higher the stage amplification at this frequency. The response characteristic of the bass boost network is shown in Fig. 7-12.

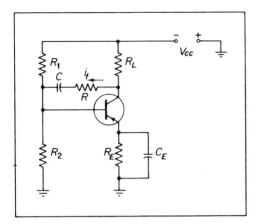

Figure 7-11 *Circuit to Provide a Low-Frequency Boost*

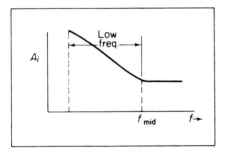

Figure 7-12 *Bass Boost Response Curve*

The network current feedback factor is given by:

$$\beta_i = \frac{i_f}{i_c} = \frac{R_L}{R_L + R - jX_c}$$

TREBLE BOOST

The basic circuit of a treble or high frequency boost is shown in Fig. 7-13. Note that at midfrequencies the reactance of X_c is relatively high resulting in a larger feedback current i_1. As frequency is increased, X_c is decreased and feedback current i_1 is decreased.

analysis Note that feedback current factor, β_i, is defined by the ratio of the feedback current with respect to the output or collector current. Thus,

$$\beta_i = \frac{i_1}{i_c}$$

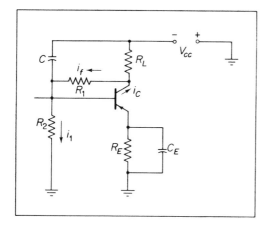

Figure 7-13 *Circuit for Treble Boost*

Using a piecewise approach, the ratio of the current i_f to i_c is given by:

$$\frac{i_f}{i_c} = \frac{R_L}{R_L + R_1 + \dfrac{R_2(-jX_c)}{R_2 - jX_c}}$$

Using the current divider theorem, the current i_1 is given by:

$$i_1 = \frac{i_f(-jX_c)}{R_2 - jX_c}$$

Consequently, with substitution and simplification, the resultant equation for the current feedback factor β_i is given by:

$$\beta_i = \frac{R_L(-jX_c)}{(R_1 + R_L)(R_2 - jX_c) + R_2(-jX_c)}$$

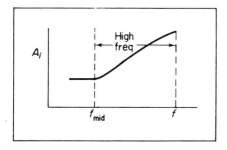

Figure 7-14 *Response Curve for a Treble Boost Circuit*

The typical response curve for a treble boost circuit is shown in Fig. 7-14.

COMBINED BASS AND TREBLE BOOST CIRCUIT

One of the circuit requirements may be to provide both bass and treble boost simultaneously. A typical circuit that can be used for this purpose is shown in Fig. 7-15.

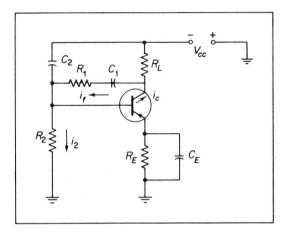

Figure 7-15 *Circuit to Provide Bass and Treble Boost*

Note that at low frequencies the reactance of X_{c_2} is so large that the actual feedback network is provided by R_2, R_1 and C_1. At the high frequencies, the reactance of X_{c_1} is extremely small so that the feedback network is established by R_1, R_2 and C_2.

analysis The equivalent circuit of the amplifier is shown in Fig. 7-16. The

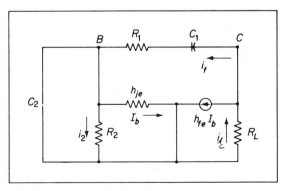

Figure 7-16 *Equivalent Circuit of Combination Feedback*

total feedback impedance is determined by:

$$Z_{\text{feedback}} = R_1 - jX_{c_1} + \frac{R_2(-jX_{c_2})}{R_2 - jX_{c_2}}$$

The ratio of the feedback current to the output or collector current is given by:

$$\frac{i_f}{i_c} = \frac{R_L}{R_L + R_1 - jX_{c_1} + \dfrac{R_2(-jX_{c_2})}{R - jX_{c_2}}}$$

The required feedback factor, β_i, is given by the ratio of the feedback current i_2 to the output or collector current. The current i_2 is determined using the current divider theorem. Thus,

$$i_2 = \frac{i_f(-jX_{c_2})}{R - jX_{c_2}}$$

The value of β_i is then given by:

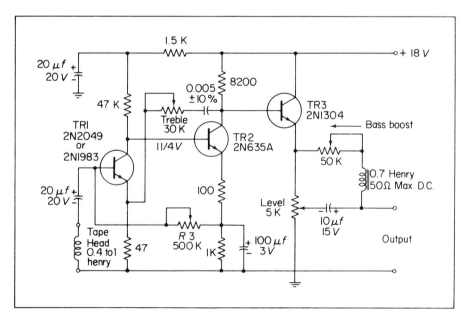

Figure 7-17 *Commercial Tape Preamplifier (Courtesy of General Electric Company)*

$$\beta_i = \frac{R_L(-jX_{c_2})}{(R_L + R_1 - jX_{c_1})(R_2 - jX_{c_2}) + R_2(-jX_{c_2})}$$

A typical commercial amplifier utilizing bass and treble boost circuits is shown in Fig. 7-17. According to the manufacturer, this particular circuit will provide a 55 db signal to noise ratio. A signal of 1.2 millivolts produces a 1 volt signal output at a midfrequency of 1 kHz. For further information concerning this particular tape amplifier, all inquiries should be directed to General Electric Company.

problems

1. An RC coupled amplifier has the following given data:
   $$A_{v_{mid}} = -100 \qquad f_1 = 200 \text{ Hz} \qquad f_2 = 30 \text{ kHz}$$
   A feedback circuit with $\beta = 0.1$ is incorporated.
   Find: (a) A_{v_r} (b) f'_1 (c) f'_2

2. An RC coupled amplifier has the following given data:
   $$A_{v_{mid}} = -50 \qquad f_1 = 150 \text{ Hz} \qquad f_2 = 40 \text{ kHz}$$
   A feedback network with $\beta = 0.2$ is incorporated.
   Find: (a) A_{v_r} (b) f'_1 (c) f'_2

3. An amplifier with feedback has a voltage amplification of 5 and a bandwidth of 10 Hz to 1.6 MHz. Determine the midfrequency voltage amplification and bandwidth without feedback ($\beta = .19$).

4. An amplifier with feedback ($\beta = .18$) has a voltage amplification of 5 and a bandwidth of 25 Hz to 1.2 MHz. Determine the voltage amplification and bandwidth without feedback.

5. For the circuit of Fig. 7-2 the following values are given:
   $$R_E = 1.5 \text{ K} \qquad R_L = 4 \text{ K} \qquad h_{ie} = 2 \text{ K} \qquad h_{fe} = 50$$
   Find: (a) A_{v_r} (b) R_{in}

6. For the circuit given in Fig. 7-2, the following values are given:
   $$R_E = 500 \ \Omega \qquad R_L = 5 \text{ K} \qquad h_{ie} = 1.5 \text{ K} \qquad h_{fe} = 70$$
   Find: (a) A_{v_r} (b) R_{in}

7. For the circuit of Fig. 7-3, the following values are given:
   $$A_{v_{mid}} = -100 \qquad r_d = 40 \text{ k}\Omega \qquad R_s = 1 \text{ k}\Omega \qquad R_L = 10 \text{ k}\Omega$$
   Find: (a) A_{v_r} (b) r_{d_r}

8. For the circuit of Fig. 7-3, the following values are given:
   $$r_d = 75 \text{ k}\Omega \qquad R_L = 30 \text{ k}\Omega \qquad A_{v_r} = -10$$
   Determine the value of R_s if the value of $A_{v_{mid}}$ is equal to -60.

9. For the circuit of Fig. 7-3, the following values are given:
   $$r_d = 50 \text{ k}\Omega \qquad R_L = 25 \text{ k}\Omega \qquad R_s = 1.5 \text{ k}\Omega \qquad u = 60$$
   Find: (a) A_{v_r} (b) r_{d_r}

10. For the circuit shown in Fig. 7-4, the following values are given:
    $$R_L = 5 \text{ k}\Omega \qquad R_F = 50 \text{ k}\Omega \qquad A_{i_{mid}} = -25 \qquad R_{in} = 1.75 \text{ K}$$
    Find: (a) A_{i_r} (b) R_{in_r}

11. For the circuit shown in Fig. 7-4, the following data are given:
    $$R_L = 6 \text{ k}\Omega \qquad A_{i_{mid}} = -30 \qquad R_{in_r} = 150 \ \Omega \qquad R_{in} = 1.5 \text{ k}\Omega$$
    Find: (a) A_{i_r} (b) R_F

12. Given the circuit shown in Fig. 7-6, the following data are known:

    $r_d = 50\text{ k}\Omega \quad R_L = 75\text{ k}\Omega \quad R_1 = 200\text{ k}\Omega \quad R_2 = 100\text{ k}\Omega$
    $u = 50$
    Find: (a) A_{v_r} (b) β

13. Given the circuit shown in Fig. 7-6, the following data are known:

    $R_L = 100\text{ k}\Omega \quad R_1 = 100\text{ k}\Omega \quad R_2 = 20\text{ k}\Omega \quad r_d = 40\text{ k}\Omega$
    $A_{v_r} = -5$
    Find: (a) $A_{v_{mid}}$ and u (b) u_r and r_{d_r}

14. Given the circuit shown in Fig. 7-8, and the following data:

    $R_L = 50\text{ k}\Omega \quad R_1 = 150\text{ k}\Omega \quad R_2 = 50\text{ k}\Omega \quad r_d = 50\text{ k}\Omega$
    $u = 70 \quad R_s = 5\text{k}\Omega$
    Find: (a) β_1 and β_2 (b) A_{v_r} (c) u'_r and r'_{d_r}

15. Given the circuit shown in Fig. 7-8 and the following data:

    $R_L = 40\text{ k}\Omega \quad R_1 = 100\text{ k}\Omega \quad R_2 = 40\text{ k}\Omega \quad r_d = 60\text{ k}\Omega$
    $u = 100 \quad R_s = 2\text{ k}\Omega$
    Find: (a) β_1 and β_2 (b) A_{v_r} (c) u'_r and r'_{d_r}

eight

SINUSOIDAL

OSCILLATORS

INTRODUCTION

A sinusoidal oscillator may be defined as a circuit that delivers alternating current output voltage and frequency *without the use of an external signal*. Essentially, the sinusoidal feedback oscillator is an amplifier that derives its input signal from its own output.

Feedback circuits of many types are available. These may have different physical elements, but each oscillator circuit must contain all the necessary requirements to sustain itself in oscillation. These conditions are:

The feedback voltage from the signal output to the signal input must be such that the value of $A\beta$ must equal unity at an angle of zero. That is:

(a) Real part of Aβ must equal unity.

(b) Imaginary part of Aβ must equal zero.

These two conditions yield the frequency of oscillation and the required value of the circuit parameter necessary to start and sustain the system in oscillation.

theory The application of dc power to the circuit causes an initial disturbance that starts the operation. A voltage resulting from this disturbance is fed back to the input and appears as an amplified output. This sequence is repetitive until the amplifier operates in the saturation and cutoff regions of the amplifying device. Eventually, the amplifier amplitude becomes steady state at some desired frequency. The variation to the outer amplifier limits that exist between saturation and cutoff must be filtered by a high Q resonant load circuit. The resonant circuit for determining the operating frequency is usually a simple coil and capacitor or an equivalent electromechanical resonator (crystal).

LC OSCILLATORS

The basic *LC* oscillator circuit is shown in Fig. 8-1. The action of the *LC* circuit is analyzed using a step by step procedure.

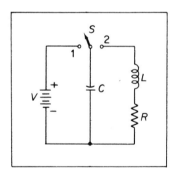

Figure 8-1 *LC Circuit*

procedure

Step 1: Set switch S to position 1. The capacitor C is charged to the applied voltage as shown in Fig. 8-2(a).

Step 2: Capacitor C is fully charged. Place switch S to position 2. Current i starts to flow as capacitor C begins to discharge through L. Due to the inductance L, the current builds slowly toward its maximum value. Maximum current and,

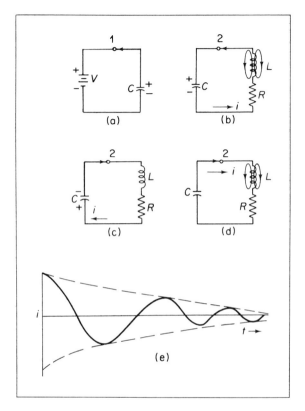

Figure 8-2   *Damped Oscillations in an LC Circuit*

consequently, the maximum magnetic field occurs when the capacitor is fully discharged. This condition is shown in Fig. 8-2(b).

Step 3:   Once the capacitor is fully discharged the magnetic field starts to collapse, but the counter emf will keep current flowing in the same direction charging the capacitor with opposite polarity, as shown in Fig. 8-2(c).

Step 4:   As the capacitor charge increases, the current magnitude decreases and the magnetic field collapses. In the ideal case, the magnetic field drops to zero when the capacitor charges to the original or applied voltage (E).

Step 5:   The process now reverses again. Capacitor C starts to discharge with the current flowing in the opposite direction.

Fig. 8-2(d) illustrates the resultant effect when capacitor C is fully discharged and maximum current and magnetic field exist.

Step 6: This interchange or *oscillation* of energy is repetitive and if the LC circuit were without loss (no resistance) would produce a perfect sine wave. Since resistance is always present, the amplitude of oscillation will eventually die out or reduce to zero as shown in Fig. 8-2(e). Although the amplitude of the sine wave may vary, the frequency of oscillation is usually determined by:

$$f_0 = \frac{1}{2\pi\sqrt{LC}}$$

Consider the circuit shown in Fig. 8-3. This type of circuit is known as a tickler feedback or tuned collector oscillator. In this case, the polarity of the coil windings is such that the base voltage will reverse an additional 180 degrees to maintain the conditions required for oscillation. The selection of the proper operating point is necessary to provide a dynamic load line. The impedance of the tuned circuit at resonance controls the dynamic load line and determines the amplitude of oscillation. The frequency of oscillation is always controlled primarily by the tuned circuit. Oscillator circuits must be analyzed to determine the frequency of oscillation and also to determine the transistor requirements necessary to sustain the system in oscillation.

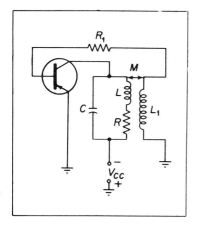

Figure 8-3 *BJT Tuned Collector Oscillator*

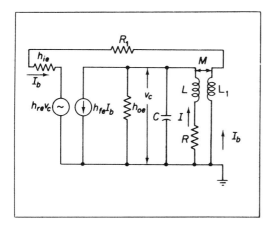

Figure 8-4 *Equivalent Circuit*

The equivalent circuit used for system analysis is shown in Fig. 8-4.

analysis The equation for the voltage amplification of the circuit without feedback is:

$$A_v = \frac{-h_{fe}Z_L}{h_{ie} + (\Delta h)\, Z_L}$$

where:

$$(\Delta h) = h_{ie}h_{oe} - h_{re}h_{fe}$$

The output voltage across the tuned circuit is given by the relationship:

$$V_o = -I(R + j\omega L)$$

The feedback voltage or the voltage induced in the secondary is given by:

$$V_{fb} = I(j\omega M)$$

The ratio of V_{fb} to V_o defines the feedback factor β. Thus,

$$\beta = \frac{-j\omega M}{R + j\omega L}$$

Note that the resultant voltage amplification of the system is given by:

$$A_{v_r} = \frac{A_v}{1 - A_v \beta}$$

and for oscillation to exist, $1 - A_v\beta = 0$. Substituting into this equation yields:

$$1 - \left(\frac{-j\omega M}{R + j\omega L}\right)\frac{-h_{fe}Z_L}{h_{ie} + (\Delta h)\, Z_L} = 0$$

Note that Z_L is defined as:

$$Z_L = \frac{R + j\omega L}{1 - \omega^2 LC + j\omega RC}$$

Substituting and simplifying yields:

$$1 = \frac{j\omega M h_{fe}}{h_{ie}(1 - \omega^2 LC + j\omega RC) + (\Delta h)(R + j\omega L)}$$

Setting real terms equal:

$$h_{ie}(1 - \omega^2 LC) + (\Delta h)R = 0$$

and

$$\omega_{osc} = \frac{1}{\sqrt{LC}}\left(1 + \frac{(\Delta h)R}{h_{ie}}\right)$$

Setting the imaginary terms equal:

$$h_{ie}R\omega C + (\Delta h)\omega L = \omega M h_{fe}$$

and

$$M = \frac{h_{ie}RC + (\Delta h)L}{h_{fe}}$$

Note that the frequency of oscillation depends on the tuned circuit values as well as the transistor parameters. Note that if (Δh) is small, the frequency of oscillation is specified by the tuned circuit. The equation for M necessary to sustain the system in oscillation is dependent on the h_{fe} as well as the transistor parameters.

TUNED DRAIN OSCILLATOR (FET)

A number of oscillator circuits have been developed and these possess varying characteristics. The high input impedance and high voltage amplification of the FET lends itself to simplicity and efficiency in many oscillator circuits. Fig. 8-5 illustrates the basic circuit of a tuned drain oscillator.

analysis The analysis of the tuned drain oscillator is similar to the tuned collector oscillator. The voltage amplification of the system without feedback

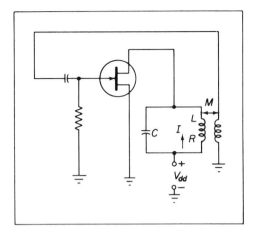

Figure 8-5 *Tuned Drain Oscillator*

is given by:

$$A_v = \frac{-uZ_L}{r_d + Z_L}$$

where:

$$Z_L = \frac{R + j\omega L}{1 - \omega^2 LC + j\omega RC}$$

The value of the output voltage is given by:

Again the feedback factor is given by:

$$\beta = \frac{-j\omega M}{R + j\omega L}$$

Then, $A_v\beta = 1 + j0$

and $$\frac{-u\left[\dfrac{R + j\omega L}{1 - \omega^2 LC + j\omega RC}\right]}{r_d + \left(\dfrac{R + j\omega L}{1 - \omega^2 LC + j\omega RC}\right)} \left(\frac{-j\omega M}{R + j\omega L}\right) = 1 + j0$$

This term simplifies to:

$$j\omega Mu = r_d(1 - \omega^2 LC) + j\omega r_d RC + R + j\omega L$$

Setting real terms equal yields:

$$r_d(1 - \omega^2 LC) + R = 0$$

Solving for ω_{osc} results in:

$$\omega_{osc} = \frac{1}{\sqrt{LC}}\sqrt{1 + \frac{R}{r_d}}$$

Setting the imaginary terms equal yields:

$$\omega Mu = \omega r_d RC + \omega L$$

and

$$M = \frac{r_d RC + L}{u}$$

COLPITTS OSCILLATOR

The Colpitts oscillator is shown in Fig. 8-6(a). This circuit utilizes dual tuning capacitors, namely C_1 and C_2. Note that the oscillator coil has only two terminals. The feedback signal is returned between base and ground. As the base becomes positive, the collector goes negative developing the potential polarities across capacitors C_1 and C_2, as shown in Fig. 8-6(b). The voltage developed across C_2 is fed back between base and ground. Since the negative side of C_2 is grounded, the voltage fed back is 180 degrees out of phase, thus maintaining the required phase relationship at the base for sustained oscillation.

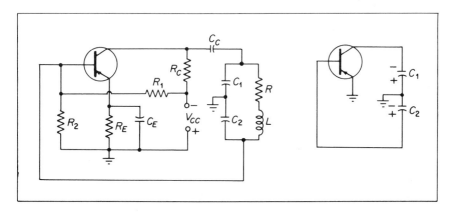

Figure 8-6 (a) *Transistor Colpitts Oscillator* (b) *Simplified Feedback Network*

Very often a split stator capacitor is used with the Colpitts oscillator. This type of capacitor has two stators and a grounded rotor. Variation of the rotor results in either an increase or decrease in both C_1 and C_2 simultaneously and is used to provide the required tuning and voltage division.

CLAPP OSCILLATOR

Modification of the Colpitts oscillator by including a capacitor in series with the coil winding results in the Clapp oscillator shown in Fig. 8-6(c). This added capacitance improves the frequency stability of the oscillator. When the total capacitance of C_1 in series with C_2 is much greater than the C_3 capacitor, then the oscillator frequency can be approximately determined by the inductor multiplied by the total series capacitance of the three capacitors. The capacitance can be made variable to provide oscillations over a wide range of frequencies.

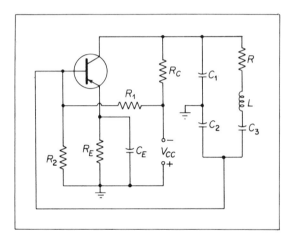

Figure 8-6   (c) *Clapp Oscillator*

HARTLEY OSCILLATOR

A Hartley oscillator is characterized by a tapped inductor in the oscillator tuned circuit. A shunt-fed Hartley oscillator is shown in Fig. 8-6(d). Resistors R_c, R_1 and R_2 are used to provide the necessary bias conditions for the circuit. The frequency determining network consists of the inductor in parallel with the capacitor, C. Since capacitor C is variable, the circuit may be tuned over a wide range of frequencies. Note that the function of C_c is to block dc and provide an ac path from the collector to the tuned circuit. Capacitor C_E provides an ac bypass around the bias resistor R_E.

A series-fed Hartley oscillator is shown in Fig. 8-6(e). The shunt-fed and series-fed Hartley oscillators are operationally identical, but differ in the method of obtaining collector bias.

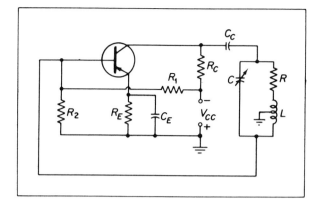

Figure 8-6 *(d) Shunt-fed Hartley Oscillator*

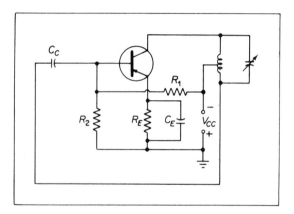

Figure 8-6 *(e) Series-fed Hartley Oscillator*

TUNED CIRCUIT OSCILLATORS

Fig. 8-7 illustrates the basic circuit of a feedback oscillator circuit utilizing a reactance element to supply feedback from the output to the input circuit.

The tuned circuit supplies a 180-degree phase shift between the output collector voltage and the voltage returned to the base. Note that the common emitter circuit supplies another 180-degree phase shift between collector and base voltages and in this way meets the requirement for oscillation.

Oscillator operation is usually Class C, but the application of the equivalent circuit theory is justified on the basis of simplified analysis and ease of understanding.

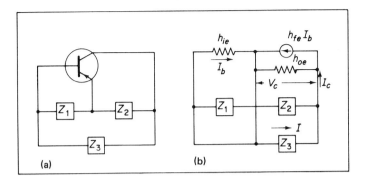

Figure 8-7 (a) Basic Oscillator Circuit (b) Equivalent Circuit

analysis The assumption made according to the equivalent circuit is that h_{re} is assumed negligible. Then, using a current summation from the equivalent circuit, we get

$$v_c = (I - I_c)Z_2 \qquad I_c = v_c h_{oe} + h_{fe}I_b$$

Combining yields:

$$v_c = (I - h_{fe}I_b)\left(\frac{Z_2}{1 + h_{oe}Z_2}\right)$$

$$v_c = -I\left[Z_3 + \frac{Z_1 h_{ie}}{Z_1 + h_{ie}}\right]$$

Solving for the transistor system current amplification yields:

$$\frac{I}{I_b} = \frac{h_{fe}Z_2}{(1 + h_{oe}Z_2)\left(Z_3 + \frac{Z_1 h_{ie}}{Z_1 + h_{ie}} + \frac{Z_2}{1 + h_{oe}Z_2}\right)}$$

The value of I_b is determined by:

$$I_b = I\left(\frac{-Z_1}{Z_1 + h_{ie}}\right)$$

Consequently, the feedback factor β is given by:

$$\beta = \frac{I_b}{I} = \frac{-Z_1}{Z_1 + h_{ie}}$$

Since $A\beta = 1 + j0$, then

$$\left(\frac{-Z_1}{Z_1 + h_{ie}}\right)\left[\frac{h_{fe}Z_2}{Z_2 + (1 + h_{oe}Z_2)\left(Z_3 + \frac{Z_1 h_{ie}}{Z_1 + h_{ie}}\right)}\right] = 1 + j0$$

This term simplifes to:

$$Z_1Z_2(1 + h_{fe}) = h_{ie}(Z_1 + Z_2 + Z_3) + h_{ie}h_{oe}Z_2(Z_1 + Z_3)$$
$$+ (Z_1Z_3)(1 + h_{oe}Z_2)$$

This equation is known as the general equation for tuned circuit oscillators. It can be applied to the Hartley, Colpitts, Clapp and tuned-collector tuned-base type oscillators.

The best known tuned circuit oscillators are named after the radio pioneers: Hartley, Colpitts and Clapp. Note that the circuits are almost identical except for the values of Z_1, Z_2 and Z_3. Thus,

Hartley	*Colpitts*	*Clapp*
$Z_1 = R_1 + j\omega L_1$	$Z_1 = \dfrac{1}{j\omega C_1}$	$Z_1 = \dfrac{1}{j\omega C_1}$
$Z_2 = R_2 + j\omega L_2$	$Z_2 = \dfrac{1}{j\omega C_2}$	$Z_2 = \dfrac{1}{j\omega C_2}$
$Z_3 = \dfrac{1}{j\omega C}$	$Z_3 = R + j\omega L$	$Z_3 = R + j(X_L - X_{c_3})$

Consider the Colpitts oscillator where the coil used is a high Q coil $(R \cong 0)$. Thus,

$$Z_1 = \frac{1}{j\omega C_1} \qquad Z_2 = \frac{1}{j\omega C_2} \qquad Z_3 = j\omega L$$

Substituting these values into the general equation and letting $C_T = \dfrac{C_1C_2}{C_1 + C_2}$ yields:

$$\frac{(1 + h_{fe})}{\omega^2 C_1 C_2} = jh_{ie}\left(\omega L - \frac{1}{\omega C_T}\right) + \frac{h_{ie}h_{oe}}{\omega C_2}\left(\omega L - \frac{1}{\omega C_1}\right) + \frac{L}{C_1}\left(1 + \frac{h_{oe}}{j\omega C_2}\right)$$

Setting the imaginary terms equal to zero yields:

$$0 = h_{ie}\left(\omega L - \frac{1}{\omega C_T}\right) - \frac{Lh_{oe}}{\omega C_1 C_2}$$

Solving for ω_{osc}:

$$\omega_{osc} = \frac{1}{\sqrt{LC_T}}\sqrt{1 + \frac{h_{oe}L}{h_{ie}(C_1 + C_2)}}$$

In most practical cases, the term $\dfrac{h_{oe}L}{h_{ie}(C_1 + C_2)}$ is much smaller than unity, consequently the frequency of oscillation is generally determined by:

$$\omega_{\text{osc}} = \frac{1}{\sqrt{LC_T}}$$

Setting the real terms equal to zero yields:

$$\frac{h_{fe} + 1}{\omega^2 C_1 C_2} = \frac{h_{ie}h_{oe}(\omega^2 LC_1 - 1)}{\omega^2 C_1 C_2} + \frac{L}{C_1}$$

Assuming that $\left\{\begin{aligned} \omega^2 &\cong \frac{1}{LC_T} \\ h_{oe} &\cong 0 \end{aligned}\right\}$, the solution for $h_{fe_{\text{required}}}$ is:

$$h_{fe_{\text{req}}} = \frac{C_2}{C_1}$$

FET TUNED CIRCUIT OSCILLATORS

The basic circuit of a tuned-feedback oscillator using an FET is shown in Fig. 8-8.

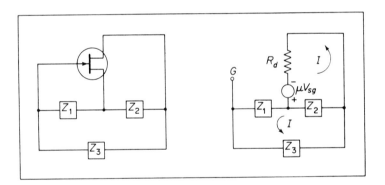

Figure 8-8 (a) *Basic Feedback Oscillator* (b) *Equivalent Circuit*

analysis The open circuit voltage amplification is given by:

$$A_v = \frac{-uZ_L}{r_d + Z_L} \quad \text{where } Z_L = \frac{Z_2(Z_1 + Z_3)}{Z_1 + Z_2 + Z_3}$$

The feedback factor β is given by:

$$\beta = \frac{Z_1}{Z_1 + Z_3}$$

The product of $A_v\beta$ is equal to:

$$\left(\dfrac{-u\dfrac{Z_2(Z_1+Z_3)}{Z_1+Z_2+Z_3}}{r_d+\dfrac{Z_2(Z_1+Z_3)}{Z_1+Z_2+Z_3}}\right)\left(\dfrac{Z_1}{Z_1+Z_3}\right)=A_v\beta$$

Since $A_v\beta$ must equal $1+j0$ for oscillation to be sustained, the following equation results.

$$r_d(Z_1+Z_2+Z_3)+(1+u)Z_1Z_2+Z_2Z_3=0$$

This resultant equation is the general equation for the Colpitts, Hartley, Clapp and tuned-drain tuned-gate type oscillators.

The technique for the use of the general equation is demonstrated in the solution to the Colpitts oscillator given in the sample problem:

sample problem

The Colpitts oscillator has the following given circuit elements:

$$Z_1=\frac{1}{j\omega C_1} \qquad Z_2=\frac{1}{j\omega C_2} \qquad Z_3=R+j\omega L$$

Solve for ω_{osc} and the u_{req} to sustain the system in oscillation.

Solution:

Step 1: Substitute the element values into the general equation:

$$r_d\left(\frac{1}{j\omega C_1}+\frac{1}{j\omega C_2}+R+j\omega L\right)+\frac{(1+u)}{-\omega^2C_1C_2}+\frac{R}{j\omega C_2}+\frac{L}{C_2}=0$$

Step 2: Set the imaginary terms equal to zero:

$$r_d\left(\omega L-\frac{1}{\omega C_T}\right)-\frac{R}{\omega C_2}=0$$

$$\text{where } C_T=\frac{C_1C_2}{C_1+C_2} \quad \text{and}$$

$$\omega_{osc}=\sqrt{\frac{1}{LC_T}}\sqrt{1+\frac{RC_T}{r_dC_2}}$$

Step 3: Set the real terms equal to zero:

$$r_dR-\frac{1+u}{\omega^2C_1C_2}+\frac{L}{C_2}=0$$

The solution for u_{req} is given by:

$$u_{req}=\left(Rr_d+\frac{L}{C_2}\right)\omega^2C_1C_2-1$$

Note that ω_{osc} is required before u_{req} can be solved.

A typical Colpitts oscillator using a BJT is shown in Fig. 8-9.

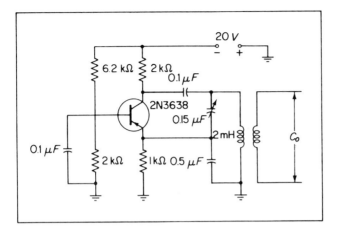

Figure 8-9 *Colpitts Oscillator Operating at About 20 kHz*

CRYSTAL OSCILLATORS

Crystals used for controlling the frequency of oscillation must have extremely good frequency stability. Such crystals are sections cut from a piezoelectric or quartz crystal with the flat sides normal to the electrical axis. The proper excitation will produce an output frequency that is a direct function of the crystal dimensions, the method of crystal cut, the method of crystal mounting and the type of holder. The symbol and equivalent circuit of the crystal and its holder are shown in Fig. 8-10.

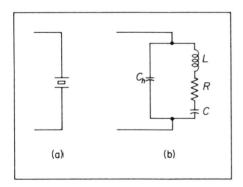

Figure 8-10 *Piezoelectric Crystal (a) Symbol (b) Equivalent Circuit*

The various electrical analogues represent the mechanical parameters of the crystal. Thus, L is the mechanical vibrating mass, C is the mechanical compliance, R is the damping resistance and C_h represents the electrical capacitance of the holder. The series resonance of the electrical circuit corresponds to mechanical resonance. The Q of the crystal is extremely high with values ranging into the thousands. The resonant frequency of the crystal may be determined by the mathematical relationship:

$$f_r = \frac{1}{2\pi\sqrt{LC_T}}$$

where:

$$C_T = \frac{CC_h}{C + C_h}$$

In practical circuits, crystals, would be mounted in an oven to maintain a preset temperature to reduce frequency drift. Oscillator circuits incorporating crystals appear in many forms. The circuits in Fig. 8-11 are typical oscillator circuits involving crystals.

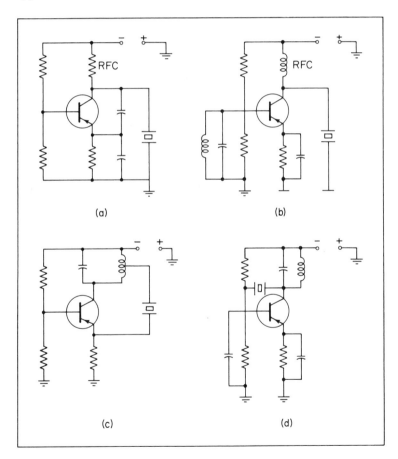

Figure 8-11 *Crystal Oscillators (a) Colpitts (b) Tuned Collector Tuned Base (c) Hartley (d) Pierce*

TROUBLESHOOTING ANALYSIS

The various possibilities that can cause loss of oscillation are: (a) defective transistor, (b) component failure in the tuned circuit, blocking capacitors or biasing circuit.

In this case, test equipment must be carefully checked to determine whether there is excessive voltage to the transistor by using a common ground on both the transistor and test chassis. High impedance meters should be used to avoid placing a dc shunt or return path in the circuit causing improper current flows and voltage distributions.

reduced output Instability should be resolved into either a *frequency* or *amplitude* instability category. Thus, *frequency instability* can occur because of:

1. Faulty soldering connections to the tuned circuit.
2. Changes in L or C in the tuned circuit. This will yield an incorrect output frequency.
3. Reduction in supply voltage. Since the operating point will change, this change causes a change in the internal capacitance of the transistor which shunts the tuned circuit.

Amplitude instability will occur because of:

1. Excessive bias produced by an increase in the emitter resistor.
2. An open emitter bypass capacitor.
3. Temperature changes varying the point of operation.
4. A leaky feedback capacitor.

PHASE SHIFT OSCILLATORS

Tuned circuits are not an essential requirement for oscillation. Oscillation merely necessitates that feedback of proper phase and magnitude is incorporated into the circuit. Several frequency sensitive networks utilizing RC networks can be applied to oscillators.

A simple transistor oscillator that utilizes RC elements to provide a frequency sensitive network with the requisite total 180-degree phase shift is shown in Fig. 8-12.

A rapid approximation would assume that each section introduces a phase shift of 60 degrees with $R_1C_1 = R_2C_2 = R_3C_3 = RC$. Consequently, it is evident that

$$\tan 60° = \sqrt{3} = \frac{1}{\omega RC}$$

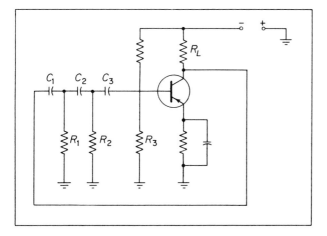

Figure 8-12 *Phase Shift Oscillator*

and

$$\omega_{\text{osc}} = \frac{1}{\sqrt{3}\,RC}$$

A more exact method utilizes the equivalent circuit shown in Fig. 8-13.

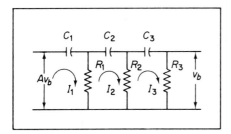

Figure 8-13 *Equivalent Circuit*

analysis The three basic equations are:

$$Av_b = I_1\left(R_1 + \frac{1}{j\,\omega C_1}\right) - I_2R_1 + 0$$

$$0 = -I_1R_1 + I_2\left(R_1 + R_2 + \frac{1}{j\,\omega C_2}\right)^{*} - I_3R_2$$

$$0 = 0 - I_2R_2 + I_3\left(R_2 + R_3 + \frac{1}{j\,\omega C_3}\right)$$

For simplicity of solution, let

$$R_1 = R_2 = R_3 = R$$
$$C_1 = C_2 = C_3 = C$$

then solving for I_3 yields:

$$I_3 = \frac{A_v v_b R^2}{\left(R + \frac{1}{j\omega C}\right)\left(2R + \frac{1}{j\omega C}\right)^2 - R^2\left(2R + \frac{1}{j\omega C}\right) - R^2\left(R + \frac{1}{j\omega C}\right)}$$

The output voltage is given by:

$$v_b = \frac{A_v v_b R^3}{R^3 + \frac{6R^2}{j\omega C} + \frac{5R}{(j\omega C)^2} + \frac{1}{(j\omega C)^3}}$$

Setting the imaginary terms equal results in:

$$-\frac{6}{\omega RC} + \frac{1}{(\omega RC)^3} = 0$$

Solving for ω_{osc} yields:

$$\omega_{osc} = \frac{1}{\sqrt{6}\,RC}$$

Setting the real terms equal to zero yields:

$$1 - A_v - \frac{5}{\omega^2 R^2 C^2} = 0$$

Substituting for ω_{osc} and solving for A_v yields:

$$A_v = -29$$

To change the frequency of oscillation, it is necessary to change the value of any element in the feedback network. For relatively large changes in frequency the three capacitors should be changed simultaneously. In all cases of frequency changes it must be remembered that the condition $A\beta = 1$ must be satisfied. Generally, because of circuit complexity, the system is not suited to variable frequency usage.

WIEN BRIDGE OSCILLATOR

A tunable circuit that has extensive usage in relatively low frequency oscillators is shown in Fig. 8-14. The two-stage circuit incorporates both positive and negative feedback; positive feedback for oscillation and negative

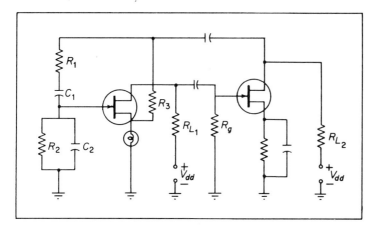

Figure 8-14 *Wien Bridge Oscillator*

feedback for amplitude limitation and improved waveform. In general, both C_1 and C_2 are used to vary the frequency of oscillation. Degenerative feedback is provided by R_3 in series with the lamp used as a source resistor. At low output, the current flow through the lamp is small and the resistance is low. As the output increases, the lamp current and the resistance increases. Consequently, the amplitude and voltage amplification are stabilized with an exceedingly good waveform.

analysis The two stage amplifier may be considered as a single stage having an overall voltage amplification of $A_v \angle 360°$. Refer to the equivalent circuit shown in Fig. 8-15.

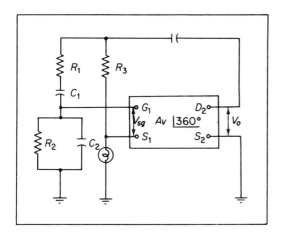

Figure 8-15 *Equivalent Circuit of Wien Bridge Oscillator*

The source to gate voltage is given by:

$$V_{sq} = \frac{V_o Z_2}{Z_1 + Z_2} - \frac{E_o R_4}{R_3 + R_4}$$

where:

$$Z_1 = R_1 + \frac{1}{j\omega C_1}$$

$$Z_2 = \frac{R_2}{1 + j\omega R_2 C_2}$$

Note also that the output voltage E_o is equal to:

$$V_o = A_v V_{sq}$$

Substituting for E_o yields:

$$\left[\frac{\dfrac{R_2}{1 + j\omega R_2 C_2}}{R + \dfrac{1}{j\omega C_1} + \dfrac{R_2}{1 + j\omega R_2 C_2}} - \frac{R_4}{R_3 + R_4} \right] A_v = 1$$

This equation simplifies to:

$$A_v \left[\frac{j\omega R_2 C_1}{1 + j\omega(R_1 C_1 + R_2 C_2 + R_2 C_1) + j^2 \omega^2 R_1 R_2 C_1 C_2} - \frac{R_4}{R_3 + R_4} \right] = 1$$

Removing the degenerative feedback term and leaving only the positive feedback terms results in a complex number set equal to zero. Thus, setting the real terms equal to zero yields:

$$1 - \omega^2 R_1 R_2 C_1 C_2 = 0$$

$$\omega_{osc} = \frac{1}{\sqrt{R_1 R_2 C_1 C_2}}$$

Setting the imaginary terms equal to zero yields:

$$\omega R_2 C_1 A_v = \omega(R_1 C_1 + R_2 C_2 + R_2 C_1)$$

$$A_v = 1 + \frac{R_1}{R_2} + \frac{C_2}{C_1}$$

If $R_1 = R_2$ and $C_1 = C_2$ then

$$\omega_{osc} = \frac{1}{RC}$$

and $A_v = 3$

A very important modification replaces the lamp by a thermistor, which is temperature dependent and nonlinear in resistance. The purpose of the thermistor is to provide an automatic control of the feedback factor β to maintain a constant gain.

NEGATIVE RESISTANCE OSCILLATORS

A PN junction diode designed for normal diode operation has an impurity concentration of 1 atom per 10^8 atoms. With this amount of doping, the width of the depletion region is approximately equal to one micron. If the concentration of the impurity atoms is increased to about 1 atom per 10^3 atoms, the diode characteristics are completely changed. This new diode was developed by Esaki. The volt ampere characteristic is shown in Fig. 8-16.

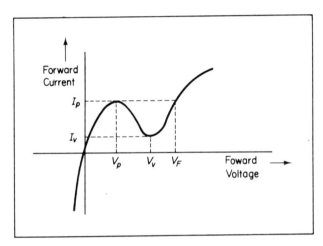

Figure 8-16   *Volt Ampere Characteristic of a Tunnel Diode*

It is evident from the volt ampere characteristic that the tunnel diode is an excellent conductor in the reverse direction. Also, for small forward voltages up to V_p, which is the peak forward voltage equal to about 50 mV, the diode resistance is low—roughly about 5 Ω. An increase of voltage from V_p to V_v causes the diode to operate in its *negative resistance region*. Operation in this region permits the diode to generate its own signals and requires merely a tuned LC circuit to select a specified frequency.

The standard symbol and equivalent circuit for the tunnel diode are shown in Fig. 8-17.

The negative resistance $(-r)$ has a minimum at the point of inflection between I_p and I_v. The series resistance R_s is the bulk ohmic resistance. The series inductance L_s is the lead inductance. The junction capacitance C

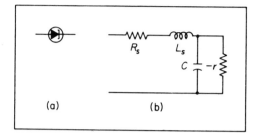

(a)   (b)

Figure 8-17   (*a*) *Symbol for Tunnel Diode* (*b*) *Small Signal Model in Negative Resistance Region*

depends on the operating point of the diode. Typical values are:

$$R_s \cong .5 \text{ to } 2 \, \Omega \qquad L_s \cong 1 \text{ to } 5 \text{ nH} \qquad C \cong 10 \text{ to } 20 \text{ pF}$$

A basic tunnel diode oscillator circuit is shown in Fig. 8-18.

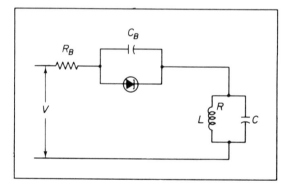

Figure 8-18   *Tunnel Diode Oscillator Circuit*

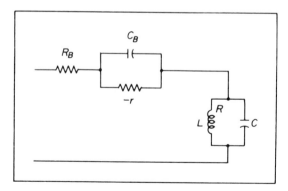

Figure 8-19   *Equivalent Circuit of the Oscillator*

The LC tuned circuit is the primary control of frequency. The equivalent circuit for the oscillator is shown in Fig. 8-19. Note that both the bulk series resistance and the lead in inductance can be neglected for most applications. The resistance R_B is used to bias the diode for proper operation. For sinusoidal output, the dc load line must intersect the negative resistance portion of the diode characteristic as shown in Fig. 8-20.

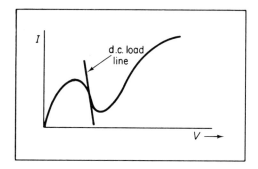

Figure 8-20   *Load Line Condition for Tunnel Diode Sinusoidal Oscillator*

For the circuit to sustain itself in oscillation, it is required that:

$$|r| = \frac{L}{CR}$$

and

$$\omega_{\text{osc}} = \frac{1}{\sqrt{LC}}$$

A typical tunnel diode sinusoidal oscillator is shown in Fig. 8-21.

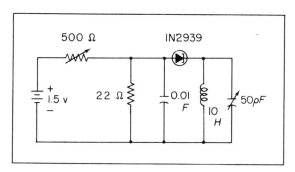

Figure 8-21   *Tunnel Diode Sinusoidal Oscillator*

problems

1. The essential features of a Hartley oscillator are shown. Assume h_{re} and h_{oe} are negligible, determine the frequency of oscillation and the h_{fe} required to sustain the circuit in oscillation.

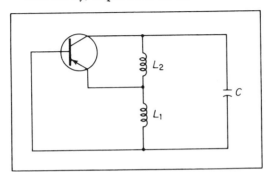

2. Determine the frequency of oscillation and the h_{fe} required to sustain the circuit in oscillation.

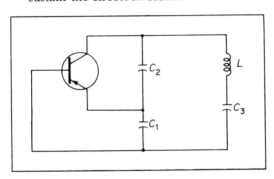

3. Given the circuit shown, determine the ω_{osc} and the value of u_{req} for the system to oscillate.

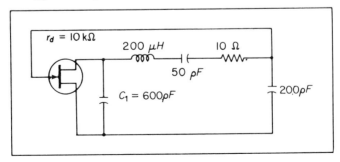

4. Given the circuit shown, determine ω_{osc} and u_{req} for the system to oscillate. $L_1 = 100\ \mu H$, $L_2 = 400\ \mu H$, $C = 200\ pF$, $r_d = 10\ K$, $Z_1 = j\omega L_1$, $Z_2 = j\omega L_2$, $Z_3 = \dfrac{1}{j\omega C}$.

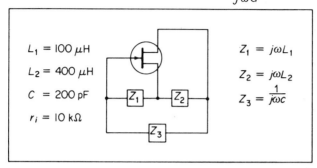

$L_1 = 100\ \mu H$

$L_2 = 400\ \mu H$

$C = 200\ pF$

$r_i = 10\ k\Omega$

$Z_1 = j\omega L_1$

$Z_2 = j\omega L_2$

$Z_3 = \dfrac{1}{j\omega C}$

5. Given the circuit shown, determine ω_{osc} and u_{req} for the system to oscillate. $C_1 = 100\ pF$, $C_2 = 500\ pF$, $L = 200\ \mu H$, $R = 10\ \Omega$, $r_d = 10\ K$, $Z_1 = \dfrac{1}{j\omega C_1}$, $Z_2 = \dfrac{1}{j\omega C_2}$, $Z_3 = R + j\omega L$.

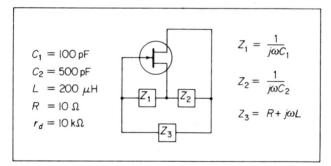

$C_1 = 100\ pF$

$C_2 = 500\ pF$

$L = 200\ \mu H$

$R = 10\ \Omega$

$r_d = 10\ k\Omega$

$Z_1 = \dfrac{1}{j\omega C_1}$

$Z_2 = \dfrac{1}{j\omega C_2}$

$Z_3 = R + j\omega L$

6. Given the circuit shown, determine ω_{osc} and u_{req} for the system to oscillate. $Z_1 = \dfrac{1}{j\omega C}$, $Z_2 = \dfrac{1}{j\omega C_2}$, $Z_3 = R + j\left(\omega L - \dfrac{1}{\omega C_3}\right)$.

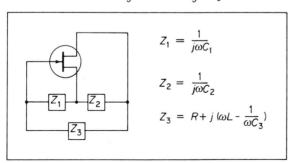

$Z_1 = \dfrac{1}{j\omega C_1}$

$Z_2 = \dfrac{1}{j\omega C_2}$

$Z_3 = R + j\left(\omega L - \dfrac{1}{\omega C_3}\right)$

7. Given the circuit shown, determine ω_{osc} and the u_{req} to sustain the system in oscillation.

$$L_1 C_1 = L_2 C_2, \quad Z_1 = \frac{j\omega L_1}{1 - \omega^2 L_1 C_1}, \quad Z_2 = \frac{j\omega L_2}{1 - \omega^2 L_2 C_2},$$

$$Z_3 = \frac{1}{j\omega C_3}, \quad r_d = 10 \text{ K}$$

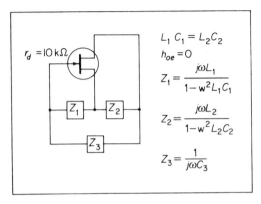

8. Given the circuit shown, determine ω_{osc} and A_{req} for the system to oscillate.

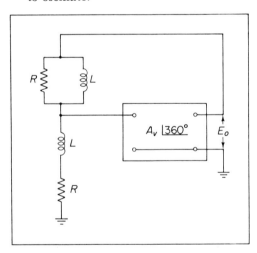

9. Given the circuit shown, determine ω_{osc} and A_{req} for the system to oscillate.

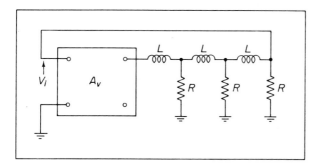

nine

MODULATION

INTRODUCTION

Signals or intelligence are transmitted through space by modulated signal carriers. Modulation is defined as the process whereby some characteristic of one waveform is varied in accordance with another waveform. Modulation is the basis of all *communications* or *radio systems*. In this system, audio signals are combined with high frequency signals that carry the audio intelligence through space. A receiver tuned to the carrier frequency is used to *pick up* the carrier and by the process of demodulation or detection recovers the original audio signals. A block diagram of a typical AM transmitter is shown in Fig. 9-1.

CHARACTERISTICS OF MODULATION

A rotating phasor with respect to time is generally described by the equation:

268

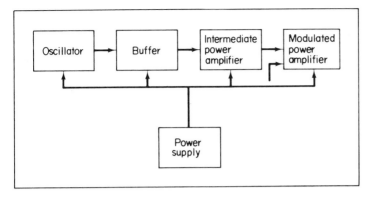

Figure 9-1   *Block Diagram of an Amplitude Modulated (AM) Transmitter*

$$v = V_m \cos (\omega t \pm \varphi)$$

The mathematical concept of a phasor representing a sinusoidal carrier rotating with a constant speed is shown in Fig. 9-2.

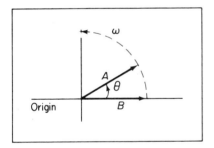

Figure 9-2   *Phasor A Leading Phasor B by 45°*

It is evident that it is possible to vary any one of the quantities in the rotating phasor. The *amplitude* of the carrier can be varied in accordance with a modulating signal. This means that the length of the modulating carrier phasor changes in accordance with the audio signal. A picture of an amplitude modulated wave (abbreviated AM) is shown in Fig. 9-3.

In phase modulation, abbreviated PM, the phase angle φ is varied while both the amplitude and carrier frequency remain constant. This type of system is shown in Fig. 9-4.

In frequency modulation, abbreviated FM, the carrier frequency ω is varied while both the phase angle and amplitude are constant.

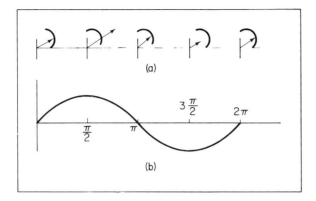

Figure 9-3   (*a*) *Amplitude Modulated Phasors* (*b*) *Input Modulating Signal*

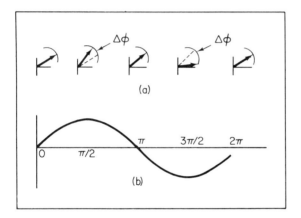

Figure 9-4   (*a*) *Angle Modulation Phasors* (*b*) *Modulating Signal*

AMPLITUDE MODULATION

In amplitude modulation (AM) the amplitude of the carrier is varied in accordance with the modulating signal. A simple example demonstrating AM is that obtained when the modulating signal *keys* the carrier, which is used in transmitting Morse code.

In this case, there are only two levels of the carrier—either on or off. This keying method is shown in Fig. 9-5.

Another method of amplitude modulation varies the amplitude of the carrier signal in accordance with the amplitude and frequency of the modulating signal. Consider the equation for the carrier amplitude signal as

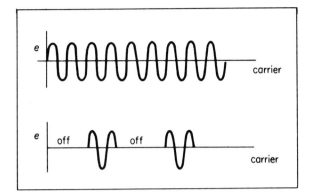

Figure 9-5   *Keying Carrier On and Off*

given by:

$$v_c = V_c \cos (\omega t + \varphi)$$

It can be assumed that the carrier frequency is generally much higher in frequency than the audio signal frequency. In addition, let φ be equal to zero. The modulating signal can be represented by the equation:

$$v_m = V_m \cos \rho t$$

Since both carrier and modulating signals are applied to an active device, the output voltage is given by:

$$v_o = (V_c + k_a V_m \cos \rho t) \cos \omega t$$

Note that the carrier amplitude is varied in accordance with both the signal amplitude and frequency. The symbol k_a is defined as a constant of proportionality. The output voltage e_o can also be written

$$v_o = V_c(1 + m_a \cos \rho t) \cos \omega t$$

where

$$m_a = \frac{k_a V_m}{E_c}$$

The term m_a is called the modulation index and varies from zero to unity. If m_a is greater than unity, overmodulation occurs with resultant distortion of signal voltages.

The expression for the amplitude modulated waveform can be put into a more useful form by using the trigonometric identity:

$$\cos A \cos B = \tfrac{1}{2} \left[\cos (A + B) + \cos (A - B) \right]$$

Substituting this identity into the voltage output equation yields the resultant

$$v = V_c \cos \omega t + \tfrac{1}{2} m_a V_c \cos (\omega + \rho)t + \tfrac{1}{2} m_a V_c \cos (\omega - \rho)t$$

The modulated output waveform is shown in Fig. 9-6.

Analysis of the waveform equation indicates that the modulated carrier consists of three components: the original carrier frequency and the

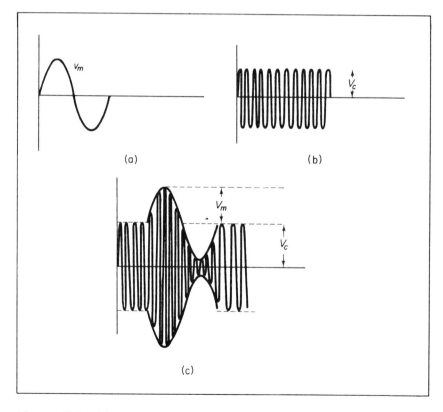

Figure 9-6   (*a*) *Modulating Signal* (*b*) *Carrier Signal* (*c*) *Amplitude Modulated Wave*

two components resulting from modulation. The frequency represented by the sum of the carrier and the modulating frequency is called the upper sideband. The frequency represented by the difference between the carrier and the modulating frequency is called the lower sideband.

In practical systems, the modulating signal is a complex waveform representing speech, music and so on. This complex waveform can be represented by an infinite series of sine and cosine waves or harmonics. Consequently, a large number of sidefrequencies will be created. The frequency spectrum illustrating the sidebands is shown in Fig. 9-7.

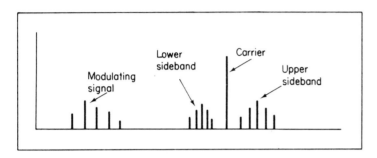

Figure 9-7 *Frequency Spectrum of AM*

The power relationships that exist when a modulated AM signal is applied to the antenna can be determined by the following mathematical relationships.

$$P_c \text{ (carrier power)} = \frac{V_c^2}{2R_{ar}} \text{ watts}$$

where

R_{ar} = characteristic impedance of the antenna.

The power due to the sidebands is given by:

$$P_{LSB} = \frac{m_a^2 V_c^2}{8R_{ar}} \text{ watts}$$

and

$$P_{USB} = \frac{m_a^2 V_c^2}{8R_{ar}} \text{ watts}$$

The total sideband power of the modulated waveform is given by:

$$P_{TSB} = \frac{m_a^2 V_c^2}{4R_{ar}} \text{ watts}$$

The total output power of the waveform is equal to the sum of the carrier and sideband powers. Thus,

$$P_o = -\frac{V_c^2}{2R_{ar}}\left(1 + \frac{m_a^2}{2}\right) \text{ watts}$$

It is evident that the power due to the sidebands is exactly equal to 50 percent of the carrier power when m_a is set equal to unity. Under these conditions, the output power is equal to 150 percent of carrier power. Thus,

$$P_o = \tfrac{3}{2}P_c \text{ watts}$$

An illustrative problem will demonstrate the theory.

sample problem

Given the equation of an amplitude modulated wave:

$$v = 100(1 + .8 \cos 3140t - .4 \cos 6280t) \cos 12.56 \times 10^6 t$$

Find: (a) *rf* components and their amplitudes.

(b) the carrier power and sideband power if $R_{ar} = 1 \text{ k}\Omega$.

(c) the crest and trough factors from the envelope.

(d) the composite modulation factor.

(e) plot the envelope for one cycle of the lower audio frequency.

Solution:

Step 1: Rewrite the voltage equation using the identity

$$\cos \alpha \cos \beta = \tfrac{1}{2}[\cos (\alpha + \beta) + \cos (\alpha - \beta)]$$

$$v = 100 \cos \omega t + 40 \cos (\omega + \rho)t + 40 \cos (\omega - \rho)t$$
$$- 20 \cos (\omega + 2\rho)t - 20 \cos (\omega - 2\rho)t$$

where:

$$\omega = 4\pi \times 10^6$$
$$\rho = \pi \times 10^3$$
$$2\rho = 2\pi \times 10^3$$

Step 2: Tabulate the amplitudes and frequencies

Frequency in MHz	Peak amplitudes
2	100
2.0005	40
1.9995	40
2.001	20
1.999	20

Step 3: Calculate the various powers

$$P_c = \frac{(100)^2}{2 \times 1 \times 10^3} = 5 \text{ W}$$

$$P_{sb_1} = \frac{2 \times 40^2}{2 \times 1 \times 10^3} = 1.6 \text{ W}$$

$$P_{sb_2} = \frac{2 \times 20^2}{2 \times 1 \times 10^3} = 400 \text{ mW}$$

$$P_{OT} = P_c + P_{sb_1} + P_{sb_2} = 7 \text{ W}$$

Step 4: Calculate the composite modulation factor

$$m_{\text{comp}} = \sqrt{m_1^2 + m_2^2}$$

$$m_{\text{comp}} = \sqrt{.8^2 + .4^2}$$

$$m_{\text{comp}} = .895$$

Step 5: Plot the equation over one cycle of audio

$$v_{env} = 100(1 + .8 \cos \rho t - .4 \cos 2\rho t)$$

ρt	e_{env}	ρt	e_{env}
0	140	210°	5.00
30°	149.5	240°	80
60°	160	270°	140
90°	140	300°	160
120°	80	330°	149.5
150°	10.7		
180°	−20.0		

Step 6: From the graph, the crest values occur at 60° and 300°. The trough value occurs at 180°.

SQUARE LAW MODULATION

One of the original methods used to produce an amplitude modulated signal used a nonlinear amplifying device and was called a square law modulator. The circuit was operated on the nonlinear portion of the static characteristic. A typical circuit as it would appear in modern technology is shown in Fig. 9-8.

Note that the carrier and audio signals are both applied to the gate input of the JFET. The modulated output voltage appears across the *drain tank circuit*. The tank circuit must have an adequate Q sufficient to pass a band of frequencies including the carrier plus and minus the audio frequencies.

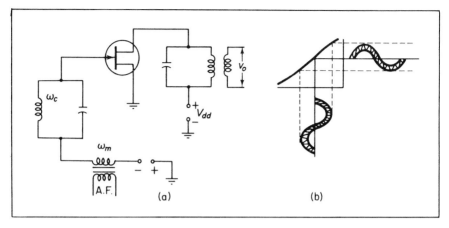

Figure 9-8   (*a*) *Square Law Modulator* (*b*) *Operating Characteristic*

analysis   The characteristic is assumed parabolic over the region of operation. The output current can be written as:

$$i_d = a_1 v_g + a_2 v_g^2$$

where a_1 and a_2 are constants. The input signal voltages are:

$$v_g = V_c \cos \omega t + V_m \cos \rho t$$

Substituting and simplifying yields the resultant:

$$i_d = a_1 V_c \cos \omega t + a_2 V_c V_m \cos (\omega + \rho)t + a_2 V_c V_m \cos (\omega - \rho)t$$

$$a_1 V_m \cos \rho t + \frac{a_2}{2} V_m^2 \cos 2\rho t + \frac{a_2}{2} V_c^2 + \frac{a_2 V_c^2}{2} \cos 2\omega t$$

The output tank circuit passes only those frequencies to which it is tuned. Thus, the output frequencies are:

ω (carrier frequency)

ω + ρ (upper sideband)

ω − ρ (lower sideband)

The output voltage can be written as:

$$v_o = i_d R_{ar}$$

and

$$v_o = [a_1 V_c \cos \omega t + a_2 V_c V_m \cos (\omega + \rho)t + a_2 V_c V_m \cos (\omega - \rho)t] R_{ar}$$

It is evident that the circuit has produced an output voltage that is an amplitude modulated signal. Circuits of this type are seldom used because of their low efficiency and low output power.

MODULATED CLASS C AMPLIFIER

A modern method of producing amplitude modulation is to vary the input power to the transistor in accordance with the modulating signal.

The operation of a transistorized Class C amplifier involves the theory and analysis of a tuned *LC* circuit similar to the one shown in Fig. 9-9.

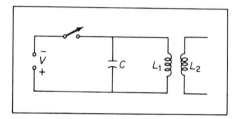

Figure 9-9 *LC Tank Circuit*

theory Apply voltage V to the *LC* tuned circuit by closing switch S; and then open switch S. Since energy has been applied to the tank circuit, an output ac signal can then be measured across the secondary winding L_2. This conversion of dc to ac power is caused by the transfer of energy from the inductance L_1 to the capacitor C at the natural resonant frequency of the circuit elements (L_1C circuit). Consequently, the introduction of dc power *shocks* the tank circuit and produces *oscillations* or *ringing* at a frequency determined by the value of L_1C combination. This ringing or oscillation will continue indefinitely providing there are no circuit losses. The losses existent in the circuit are usually the leakage resistance of the capacitor and the loss resistance in the coil. The higher the value of circuit losses, the faster the circuit oscillations will be dampened. Consequently, high-Q-tuned circuits are required for Class C operation.

Fig. 9-10 illustrates a tank circuit whose output is kept at a constant level by applying excitation input pulses in phase with the output signal.

A Class C amplifier is biased well beyond cutoff of the collector current and then the input base circuit is driven with sufficient excitation voltage to produce an input current flow for a short period of time. This drive will produce pulses in the output or collector circuit that will cause the tuned L_1C circuit to oscillate. A total conduction angle from 100 to 140 degrees is desirable for straight through operation.

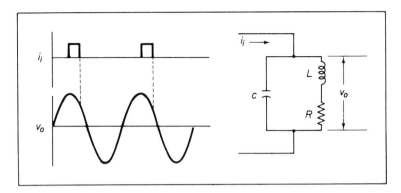

Figure 9-10   *LC Tuned Circuit at the Fundamental Frequency*

The essential features for the design of the tuned circuits in a Class C amplifier are: (a) high load impedance, (b) high Q coils must be used, (c) lossless capacitors and (d) short lead connections.

The value of impedance (R_{ar}) required by the tuned circuit is determined by:

$$R_{ar} = \frac{V_{cc}^2}{2P_{ac}} \text{ ohms}$$

An assumption is made that the peak value of the collector to emitter voltage is approximately equal to the applied dc voltage. Also note that

$$R_{ar} = QX_L$$

The loaded Q of a Class C amplifier normally varies between 10 and 15. The typical figure used is 12. Therefore, the values of L and C at the fundamental frequency of operation is given by:

$$L = \frac{R_{ar}}{12\omega}$$

and

$$C = \frac{12}{\omega R_{ar}}$$

The efficiency of the tank circuit is determined by the ratio of the loaded to the unloaded Q. The required formula is:

$$\%\eta = \frac{Q_0 - Q}{Q_0} 100\%$$

where

$$Q_0 = \text{unloaded } Q \text{ of the coil.}$$
$$Q = \text{loaded } Q \text{ of the coil.}$$

A common modification of the tuned circuit is the π network shown in Fig. 9-11.

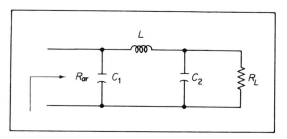

Figure 9-11 π *Network Output Circuit*

The output power is normally applied to the antenna or the resistance symbolized by R_L. The elements C_1, C_2 and L are evaluated by:
Using a Q of 12, the values are:

$$C_1 = \frac{Q}{\omega R_{ar}} = \frac{12}{\omega R_{ar}}$$

$$C_2 = \frac{1}{\omega \sqrt{\dfrac{R_{ar}R_L}{Q^2 + 1 - \dfrac{R_{ar}}{R_L}}}} = \frac{1}{\omega \sqrt{\dfrac{R_{ar}R_L}{145 - \dfrac{R_{ar}}{R_L}}}}$$

$$L = \frac{R_{ar}}{\omega(Q^2 + 1)}\left(\frac{R_{ar}}{X_{c_1}} + \frac{R_L}{X_{c_2}}\right)$$

A typical 20-watt amplitude modulated transmitter is shown in Fig. 9-12. The transmitter comprises an oscillator stage feeding a driver that in turn is fed to the modulated power output stage.

circuit description The output stage delivers 10 watts (rms) carrier power into a 50 ohm load. A capacitive tap pi-network is utilized to transform the 50 ohm load to approximately 7 ohms at the emitter. The 7 ohm emitter load is required to produce the 10 W carrier power.

The required drive power is approximately 1 W for the rated output power. The input impedance of the output transistor is 5 ohms. This input impedance is achieved by use of the toroid transformer (T_1). Any leakage and primary reactance that exist are tuned out on the primary side by the variable capacitor C across the primary.

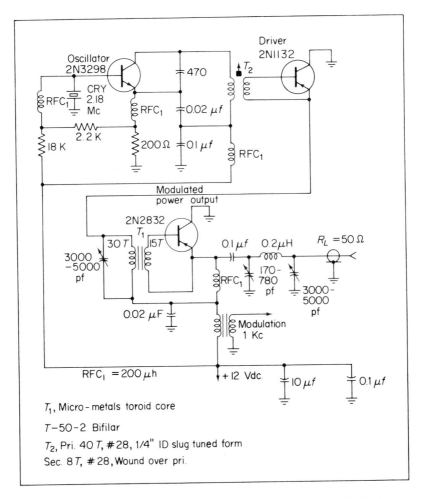

Figure 9-12 *20-Watt 2.18 Transmitter (Courtesy of Motorola, Inc.)*

The oscillator output is fed to the driver through a transformer T_2. This transformer acts as the oscillator coil and provides coupling to the driver stage. The required power level of the oscillator is approximately 40 mW. The required modulation power and resistance is 7 W and 10 ohms.

SINGLE SIDEBAND (BALANCED MODULATOR)

In the analysis of the amplitude modulated system, it was evident that the carrier was not used to convey information. All intelligence was transmitted in the side bands. If the carrier were eliminated in the output there would be an increased efficiency in transmitting intelligence. When the carrier is suppressed, and only the sidebands are transmitted, the system is called a suppressed carrier or double sideband system. Consequently, a sideband system can be devised to transmit sidebands only.

A circuit used for this purpose is called a balanced modulator and one form of this circuit is shown in Fig. 9-13.

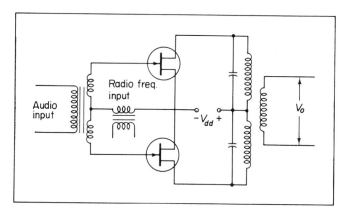

Figure 9-13   *Balanced Modulator*

A balanced modulator is a circuit with two inputs; one for a carrier and the other for the modulating signal. The output is an amplitude-modulated suppressed-carrier signal having only the lower and upper sidebands.

analysis   This type of modulation utilizes the nonlinear characteristic of the FETs so that the individual currents can be expressed by:

$$i_{d_1} = a_1 v_{g_1} + a_2 c_{g_1}^2 + \cdots$$
$$i_{d_1} = a_1 v_{g_2} + a_2 v_{g_2}^2 + \cdots$$

The two gate signals are:

$$v_{g_1} = V_c \cos \omega t + V_m \cos pt$$
$$v_{g_2} = V_c \cos \omega t - V_m \cos pt$$

The current i_{d_1} is given by:

$$i_{d_1} = a_1 V_c \cos \omega t + a_1 V_m \cos pt + a_2 V_c^2 \cos^2 \omega t + a_2 V_m^2 \cos^2 pt$$
$$+ 2a_2 V_m V_c \cos \omega t \cos pt$$

The current i_{d_2} is given by:

$$i_{d_2} = a_1 V_c \cos \omega t - a_1 V_m \cos pt + a_2 V_c^2 \cos^2 \omega t + a_2 V_m^2 \cos^2 pt$$
$$- 2a_2 V_c V_m \cos \omega t \cos pt$$

The two drain currents subtract in the push-pull transformer. The output voltage is given by:

$$v_o = j\omega M (i_{d_1} - i_{d_2})$$

or

$$v_o = j\omega M \, 4a_2 V_c V_m \cos \omega t \cos \rho t$$

The output circuit of the balanced modulator is tuned to the carrier frequency and will reject all frequencies except carrier and sidebands. The carrier is eliminated because of the push-pull operation with the result that only the sideband components are contained in the output. Thus, the output voltage is:

$$v_o = 2\omega M \, a_2 V_c V_m \left[\cos (\omega - \rho)t - \cos (\omega + \rho)t \right]$$

It should be noted that, applying the same input dc power to an AM system and a single sideband system, the peak useful output power of the single sideband transmitter is theoretically six times that of the conventional amplitude modulated system with reduced channel width.

There are essentially two methods used to produce single sideband transmission:

1. phasing of two balanced modulators
2. filter method.

Generation of a single sideband signal by the phasing method requires the use of two balanced modulators and proper phase shift networks as shown in Fig. 9-14.

analysis	It may be assumed that the balanced modulators have outputs that are in the form:

$$i = a_1 v_g + a_2 v_g^2$$

For modulator number one, the two inputs are:

$$v_{g_1} = V_m \sin \rho t + V_c \sin \omega t$$
$$v_{g_2} = -V_m \sin \rho t + V_c \sin \omega t$$

The output is then equal to:

$$V_{o_1} = 2\omega M a_2 V_c V_m \left[\cos (\omega - \rho)t - \cos (\omega + \rho)t \right]$$

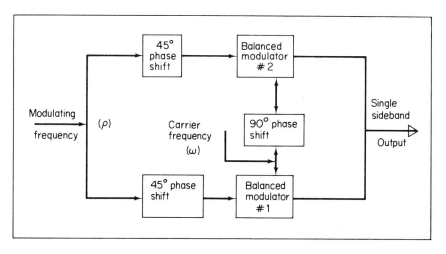

Figure 9-14 *Block Diagram Phasing Method for Producing Single Sideband*

For modulator number two, the two inputs are:

$$v'_{\sigma_1} = V_c \cos \omega t + V_m \cos \rho t$$
$$v'_{\sigma_2} = V_c \cos \omega t - V_m \cos \rho t$$

The output is then equal to:

$$V_{o_2} = 2\omega M a_2 V_c V_m [\cos (\omega - \rho)t + \cos (\omega + \rho)t]$$

If the outputs of the two modulators are directly added, the resultant equation is given by:

$$V_{.o_1} + V_{o_2} = 4\omega M a_2 V_c V_m \cos (\omega - \rho)t$$

If the outputs of the two modulators are directly subtracted, the resultant equation is given by:

$$V_{o_1} - V_{o_2} = 4\omega M a_2 V_c V_m \cos (\omega + \rho)t$$

FILTER METHOD OF SINGLE SIDEBAND TRANSMISSION

The filter method of producing single sideband is shown using the block diagram technique in Fig. 9-15.

The filter used must have the following properties: (a) sharp cutoff frequency and (b) high attenuation immediately after cutoff.

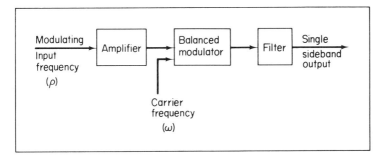

Figure 9-15 *Filter Method of Single Sideband Generation*

Filters containing high Q elements are used in single sideband generation. Quartz crystals or mechanical filters are used to provide the necessary filter operation with attenuations of 50 db for the unwanted sideband frequencies.

FREQUENCY MODULATION (FM)

Instead of changing the amplitude of the carrier in accordance with a modulating signal, the frequency may be varied in this manner. An FM modulator is a device that causes the instantaneous frequency of a carrier to vary from its unmodulated value by an amount proportional to the instantaneous amplitude of the modulating signal. The maximum frequency deviation that can be produced by the modulator determines its modulation capabilities.

An example of a modulating signal and modulated carrier is shown in Fig. 9-16.

Note that in FM the instantaneous frequency is made proportional to the modulating voltage. Let the modulating voltage be defined as e_a or:

$$v_a = V_m \cos \rho t$$

then the frequency variation is given by:

$$\omega(t) = \omega_c + k_f V_m \cos \rho t$$

where k_f is a constant of proportionality. The carrier amplitude and frequency are specified by:

$$v_c = V_{c_m} \sin \omega t$$

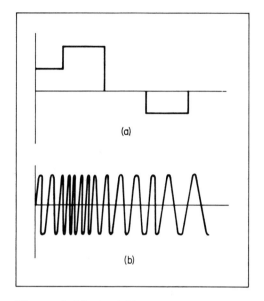

Figure 9-16 *FM (a) Modulating Signal (b) Modulated Carrier*

Substituting for ω yields the resultant equation:

$$v_c = V_{c_m} \sin\left[\omega_c t + \frac{k_f V_m}{\rho} \sin \rho t\right]$$

The instantaneous frequency is given by:

$$f = f_c \pm \frac{k_f V_m}{2\pi} \cos \rho t$$

The maximum and minimum frequencies are given by:

$$f_{\max} = f_c + \frac{k_f V_m}{2\pi}$$

$$f_{\min} = f_c - \frac{k_f V_m}{2\pi}$$

The frequency deviation f_d is defined as the maximum deviation of the carrier from its average value. Thus,

$$f_d = \frac{k_f V_m}{2\pi}$$

The modulation index (abbreviated m_f) for a frequency modulated carrier is specified by the ratio of the frequency deviation to the modulating frequency. Thus,

$$m_f = \frac{f_d}{f_a}$$

or

$$m_f = \frac{k_f V_m}{\rho}$$

Consequently, the carrier frequency and amplitude can be specified by:

$$v_c = V_{c_m} \sin (\omega_c t + m_f \sin \rho t)$$

The frequency modulated waveform is composed of a center or carrier frequency, defined by $\frac{\omega_c}{2\pi}$ and an infinite set of side frequencies, each pair spaced by an amount equal to the modulating frequency specified by ρ. Fortunately, for a modulation index equal to or greater than three, only six pairs of sidebands would be considered.

The FCC[1] establishes the maximum necessary bandwidth and the maximum allowable frequency deviation off carrier frequency. The allowable bandwidth set by the FCC is 200 kHz.

FM BY REACTANCE CIRCUIT VARIATION

In a frequency modulated transmitter the frequency deviation must be made proportional to the amplitude of the modulating signal. Methods for producing such a signal necessitate the variation of either the capacitance or inductance in the oscillator tuned circuit. One of the earliest methods utilizes a capacitor microphone placed across the tuned circuit of the oscillator. The procedure varies the capacitance of the tuned circuit at an audio rate. This was a direct method for producing FM. A more modern technique is to insert a *reactance circuit* across the tuned oscillator circuit to accomplish the desired result.

One form of the reactance circuit using FETs is shown in Fig. 9-17. In general, series RC and RL circuits are used for Z_{dg} and Z_{gs} respectively. These can be represented by the four cases shown in Fig. 9-18.

To determine whether the particular circuit acts as a capacitive or inductive reactance, phasor techniques can be used. Thus, using phasor techniques for Case I, the resultant diagram shown in Fig. 9-19 indicates that

[1]FCC denotes the Federal Communications Commission, which allocates permissible bandwidths and frequencies to networks.

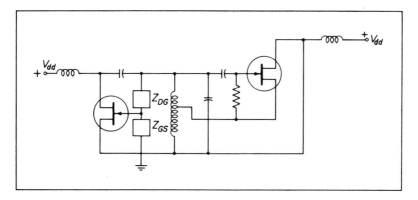

Figure 9-17   *Use of a Reactance Circuit*

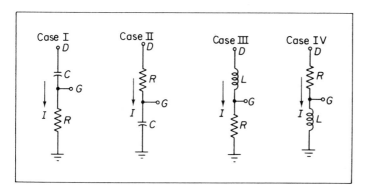

Figure 9-18   *Four Possibilities for Reactance Circuits*

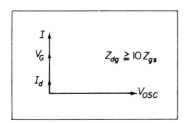

Figure 9-19   *Phasors Diagram for Case I*

the circuit acts as a capacitive reactance across the oscillator source, since current I_d leads V_{osc} by 90°.

analysis   The criterion for a reactance circuit variation is that

$$Z_{dg} \geqq 10\ Z_{gs}$$

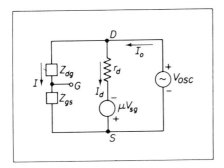

Figure 9-20 *Equivalent Circuit of Reactance Unit*

Consider the equivalent circuit shown in Fig. 9-20.

It is evident upon examination of the equivalent circuit that it can be assumed that

$$r_d \gg Z_{dg} + Z_{gs}$$

then

$$I_o = I + I_d$$

$$I = \frac{V_{osc}}{Z_{dg} + Z_{gs}}$$

$$I_d = \frac{V_{osc} + uV_{sg}}{r_d}$$

$$V_{sg} = IZ_{gs}$$

Note that the current I_o is equal to:

$$I_o = V_{osc}\left[\frac{1}{Z_{dg} + Z_{gs}} + \frac{1}{r_d} + \frac{1}{\dfrac{1}{g_m} + \dfrac{Z_{dg}}{g_m Z_{gs}}}\right]$$

It is evident that the terms r_d and $Z_{dg} + Z_{gs}$ have very little control over the oscillator frequency. The value of $\dfrac{Z_{dg}}{g_m Z_{gs}}$ determines the amount of either capacitive or inductive reactance that is presented to the oscillator tuned circuit. For example, assume Case I operation.

$$Z_{dg} = \frac{1}{j\omega C} \qquad Z_{gs} = R$$

then

$$\frac{Z_{dg}}{g_m R} = \frac{1}{j\omega(g_m RC)}$$

Therefore, $C_{eq} = g_m \, RC$. The effective capacitance is controlled directly by the g_m, which will vary with the FET selected. A graph of g_m versus v_g is shown in Fig. 9-21. The transconductance of an FET may be expressed by the assumed linear relationship

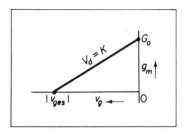

Figure 9-21   *Graph of g_m vs. v_g*

$$g_m = \frac{G_o}{|V_{g_{co}}|} v_g + G_o$$

Since $v_g = -|V_{gg}| + v_a$ and $v_a = V_a \cos pt$, then

$$g_m = \frac{G_o}{|V_{g_{co}}|} (-|V_{gg}| + V_a \cos pt) + G_o$$

The oscillator frequency with the reactance circuit across the tuned circuit is given by:

$$f_{osc} = \frac{1}{2\pi\sqrt{L_t(C_t + C_{eq})}}$$

The oscillator frequency measured with zero audio but with the reactance circuit across the oscillator is given by:

$$f_{osc} = \frac{1}{2\pi\sqrt{L_t\left[C_t + G_o RC\left(1 - \frac{|V_{gg}|}{|V_{g_{co}}|}\right)\right]}}$$

The oscillator frequency measured with audio applied is given by:

$$f_{osc} = \frac{1}{2\pi\sqrt{L_t\left[C_t + G_o RC\left(1 - \frac{|V_{gg}|}{|V_{g_{co}}|} + \frac{v_a}{|V_{g_{co}}|}\right)\right]}}$$

It is evident that the application of an audio voltage changes the g_m of the FET. The magnitude of the change in oscillator frequency resulting from the change in g_m is given by:

$$\Delta f = \frac{f_oRCG_o}{C_e|V_{g_{co}}|}\Delta V_a$$

An illustrative problem will demonstrate the theory.

sample problem

A Case I reactance circuit is to be used with an oscillator. The oscillator operates at 20 MHz with the FET in the circuit.

$$R = 5\text{ k}\Omega \qquad\qquad C_e = 200\text{ pF}$$

$$C = .0159\text{ pF} \qquad g_m = (1000 + 200\,v_a)10^{-6}\text{ mhos}$$

Determine the peak swing of the audio voltage required to produce a frequency change of 5 kHz.

Solution:

Step 1:   Note that the ratio of G_o to $|V_{g_{co}}|$ is given by 200×10^{-6}. This value is substituted into the given equation.

$$\Delta f = \frac{f_oRC200 \times 10^{-6}}{C_e}\Delta v_a$$

Step 2:

$$v_a = \frac{5 \times 10^3 \times 200 \times 10^{-12}}{200 \times 10^{-6} \times 20 \times 10^6 \times 5 \times 10^3 \times .0159(10)^{-12}}$$

$$v_a = 3.14\text{ V}$$

FM SYSTEMS

A block diagram of an FM transmitter illustrating the use of the reactance circuit method is shown in Fig. 9-22. Frequency stability of the

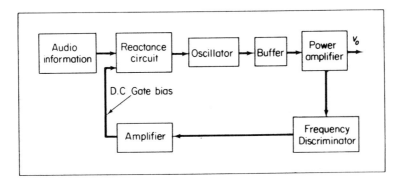

Figure 9-22   *Block Diagram of an FM Transmitter*

master oscillator is poor. Therefore, to prevent channel drift, a frequency stabilizing circuit is incorporated into the system. In this case, a portion of the output signal is removed and applied to a circuit called a frequency discriminator that produces a dc output voltage that is directly proportional to the instantaneous variation of the modulated carrier from center frequency. If channel drift occurs, the dc voltage is averaged and used to correct the reactance circuit for center frequency operation.

Generally, the output frequency is divided to a lower frequency and compared with a crystal oscillator operating at this low frequency. Any difference or error is then used to initiate action in a control motor whose shaft controls the rotation of the tuned circuit capacitor in the master oscillator circuit.

problems

1. The voltage output of a transmitter is applied to a 100 ohm load resistor.

$$v = 1000\sqrt{2}\,(1 + .2\sin \rho t)\sin \omega t \text{ V}$$

Find: the power supplied by each of the frequency components.

2. The voltage output is given by

$$v = 500\sqrt{2}\,(1 + .3\sin \rho t)\sin \omega t \text{ V}$$

and applied to a 50 ohm resistive load. Find the power supplied by each of the frequency components present.

3. Given the equation of an amplitude modulated waveform:

$$v = 50(1 + .75\cos 6280\,t - .45\cos 1{,}2560\,t)\cos 6.28(10)^6\,t$$

Find: (a) radio frequency components and their amplitudes.
(b) the power supplied by each component if $R_{ar} = 50\ \Omega$.
(c) plot the envelope for one cycle of the lower audio frequency.
(d) the composite modulation factor.

4. The following equation of an amplitude modulated waveform is given:

$$v = 10^3\sqrt{2}\,(1 + .6\sin 6280\,t - .4\sin 18840\,t)\sin 6.28 \times 10^5 t$$

Find: (a) the frequency and amplitudes of each component.
(b) $R_{ar} = 5\ \text{k}\Omega$. Find the applied power at each frequency.
(c) the composite modulation factor.
(d) plot to approximate scale one cycle of the lower audio frequency.

5. The tank circuit of a Class C amplifier must have an impedance of 2.4 kΩ at 10 MHz. Determine L and C.

6. The tank circuit of a Class C amplifier must have an impedance of 3 kΩ at 2 MHz. Determine L and C.

7. A Class C amplifier is coupled to a 50 Ω load by a π network. The impedance looking into the network is 2 kΩ. If $f = 1$ MHz and $Q = 12$, find C_1, C_2 and L.

8. A Class C amplifier is coupled to a 75 Ω load by a π network. The impedance looking into the circuit is 2.5 kΩ at 10 MHz. If $Q = 12$, determine C_1, C_2 and L.

9. Given a Case II reactance circuit, construct the phasor diagram, and prove that the value of inductance is:

$$L_{eq} = \frac{RC}{g_m}$$

10. Given a Case III reactance circuit, construct the phasor diagram and prove that the value of inductance is:

$$L_{eq} = \frac{L}{g_m R}$$

11. Given a Case IV reactance circuit, construct the phasor diagram and prove that the value of capacitance is:

$$C_{eq} = \frac{g_m L}{R}$$

12. A Case I reactance circuit is connected across an oscillator. The following data are given.
    $f_{osc} = 5$ MHz $C = .5$ pF $R = 6$ kΩ $C_{tank} = 28$ pF
    If $g_m = (4000 + 800\, v_a)10^{-6}$ mhos, determine the audio voltage necessary to produce a frequency deviation of 5 kHz.

13. A Case I reactance circuit is used. The following data are given.
    $f_{osc} = 5$ MHz with zero audio
    $C = .35$ pF $R = 5$ kΩ $L_{tank} = 50\ \mu$H
    If $a_m = (1300 + 50\, v_a)10^{-6}$ mhos, determine the audio voltage necessary to produce a frequency deviation of 15 kHz.

14. A Case IV reactance circuit is used. The following data are given. $L_{tank} = 40\ \mu$H $L = 3.18\ \mu$H $R = 15\ X_L$ $f_{osc} = 25$ MHz with zero audio. Determine the amount of audio voltage necessary to produce a frequency deviation of 20 kHz.

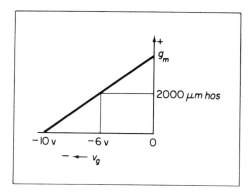

ten

DEMODULATORS

SQUARE LAW DETECTORS

When a modulated waveform containing intelligence is transmitted, some method must be employed to receive the intelligence. The process of recovering the original information is called *demodulation* or detection.

Demodulators for amplitude modulated signals can be of two types—small signal or large signal. The small signal type uses square law detection whereas the large signal type utilizes the linear characteristic of the detector.

Detectors used for FM waveforms should be designed to convert frequency variations about a center frequency into amplitude variations and are usually designed to be insensitive to amplitude variations in the signal with a resultant reduction in noise and interference.

A small amplitude modulated signal is applied to the input terminals of the square law detector shown in Fig. 10-1. Note that a diode in series with a load resistor will have the diode characteristic shown in Fig. 10-1(b) for small signal input.

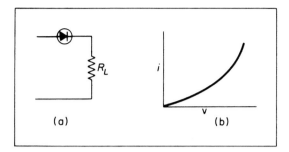

Figure 10-1   (a) *Square Law Detector*   (b) *Diode Characteristic*

The current in the circuit can be expressed by

$$i_p = a_1 v_i + a_2 v_i^2 + a_3 v_i^3 + \cdots$$

For square law detection, the first two terms of the series can be utilized and the higher order terms ignored or neglected. Thus,

$$i_p = a_1 v_i + a_2 v_i^2$$

If the modulated input signal to the detector has the equation:

$$v_i = V_{c_m}(1 + m_a \cos pt) \cos \omega t$$

the equation for the diode current, then, is:

$$i_p = a_1 V_{c_m} \cos \omega t + m_a a_1 V_{c_m} \cos pt \cos \omega t$$
$$+ a_2 V_{c_m}^2 (1 + 2m_a \cos pt + m_a^2 \cos^2 pt) \cos^2 \omega t$$

Utilizing the trigonometric identities,

$$\cos pt \cos \omega t = \tfrac{1}{2}[\cos (\omega + p)t \cos (\omega - p)t]$$
$$\cos^2 \omega t = \tfrac{1}{2}(1 + \cos 2\omega t)$$

Consequently, the expression for current is:

$$i_p = \frac{a_2 V_{c_m}^2}{2} + \frac{a_2 m_a^2 V_{c_m}^2}{4}$$
$$+ a_1 V_{c_m} \cos \omega t + \frac{a_2 V_{c_m}^2}{2}\left(1 + \frac{m_a^2}{2}\right) \cos 2\omega t$$
$$+ a_2 V_{c_m}^2 m_a \cos pt + \frac{a_2 m_a^2 V_{c_m}^2}{4} \cos 2pt$$

$$+ \frac{a_2 m_a V_{c_m}^2}{2} [\cos (2\omega - \rho)t + \cos (2\omega + \rho)t]$$

$$+ \frac{a_1 V_{c_m} m_a}{2} [\cos (\omega - \rho)t + \cos (\omega + \rho)t]$$

$$+ \frac{a_2 V_{c_m}^2 m_a^2}{8} [\cos 2(\omega - \rho)t + \cos 2(\omega + \rho)t]$$

The various frequencies existing in the output waveform are summarized in Table 10-1.

TABLE 10-1 *FREQUENCIES IN OUTPUT WAVEFORM*

Term	Frequency	Amplitude
dc	0	$\dfrac{a_2 V_{c_m}^2}{2}\left(1 + \dfrac{m_a^2}{2}\right)$
carrier	ω	$a_1 V_{c_m}$
sideband frequency	$(\omega \pm \rho)$	$\dfrac{a_1 m_a V_{c_m}}{2}$
second harmonic (carrier)	2ω	$\dfrac{a_2 V_{c_m}^2}{2}\left(1 + \dfrac{m_a^2}{2}\right)$
modulating frequency	ρ	$a_2 m_a V_{c_m}^2$
second harmonic modulating frequency	2ρ	$\dfrac{a_2 V_{c_m}^2 m_a^2}{4}$
second harmonic sidebands	$2(\omega \pm \rho)$	$\dfrac{a_2 V_{c_m}^2 m_a^2}{8}$
second harmonic carrier $\pm$ sidebands	$2\omega \pm \rho$	$\dfrac{a_2 V_{c_m}^2 m_a^2}{2}$

The output voltage that is sent to the next stage is usually the modulating signal plus the second harmonic. The amplitudes are:

fundamental modulating voltage signal amplitude $= a_2 V_{c_m}^2 m_a R \cos \rho t$

second harmonic or distortion term $= \dfrac{a_2 V_{c_m}^2 m_a^2 R \cos 2\rho t}{4}$

It is the distortion term that imposes certain limits on the use of this type of detector. Note that the magnitude of distortion will vary with respect to m_a^2. The maximum distortion is 24 percent and is completely undesirable.

LINEAR DETECTOR

A diode in series with a large resistive load will essentially have a linear characteristic as shown in Fig. 10-2.

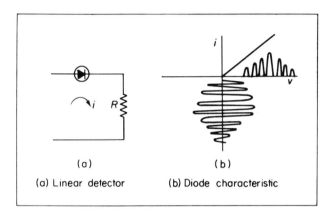

(a)   (b)

(a) Linear detector   (b) Diode characteristic

Figure 10-2 *(a) Linear Detector   (b) Diode Characteristic*

If an amplitude modulated waveform is applied, the positive portion of the signal will permit conduction, whereas the negative portion will be removed. The original signal can then be recovered as an amplitude variation about the dc or average value of the signal. The equation for the output voltage, assuming the output current is averaged over one cycle, is given by:

$$v_o = \frac{V_{c_m} R_L}{\pi (r_d + R_L)} + \frac{m_a V_{c_m} R_L}{\pi (r_d + R_L)} \cos \rho t$$

where

V_{c_m} = peak amplitude of the carrier voltage

m_a = modulation index

Note that the detector process has recovered two components. The first is the dc component, given by:

$$V_{dc} = \frac{V_{c_m} R_L}{\pi (r_d + R_L)}$$

The second is the modulation component, given by:

$$V_m = \frac{m_a V_{c_m} R_L}{\pi (r_d + R_L)} \cos \rho t$$

This elementary circuit is inefficient since the dc component is limited to approximately .33 of the input peak voltage. The circuit efficiency can be improved by inserting a filter capacitor across the load as shown in Fig. 10-3.

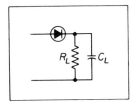

Figure 10-3   *Improved Linear Detector*

The theory of operation is quite similar to an *RC* filter in series with the rectifier. Capacitor C_L charges to the peak value of the positive signal input and discharges through R_L on the negative half cycles. The voltage across C_L and the load obviously follow the envelope of the modulated wave. This operation is shown in Fig. 10-4.

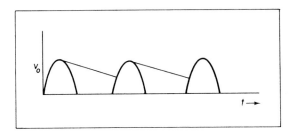

Figure 10-4   *Charge and Discharge of an RC Filter*

analysis   An amplitude modulated voltage is applied to the input terminals of the envelope detector. Thus,

$$v = V_c \left(1 + m_a \cos pt\right) \cos \omega t$$

The diode voltage will be biased by V_o, as is shown in Fig. 10-5, and the current is discontinuous.

The voltage across the diode is given by:

$$v_d = V_{c_m} \left(1 + m_a \cos pt\right) \cos \omega t - V_o$$

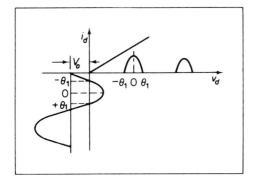

Figure 10-5 *Operation of a Linear Diode*

For simplicity of analysis, the following definitions will be used:

$$V' = V_{c_m} (1 + m_a \cos pt)$$

$$\theta = \omega t$$

Using the simplifying assumptions and definitions, the diode voltage becomes:

$$V_d = V' \cos \theta - V_o$$

The value of diode current is then equal to:

$$i_d = g_d(V' \cos \theta - V_o)\Big|_{-\theta_1}^{\theta_1}$$

Since the term V' is practically constant for a few cycles of carrier frequency, the current can be averaged by the use of higher mathematics to be equal to:

$$I_o = \frac{g_d}{\pi}(V' \sin \theta_1 - V_o\theta_1)$$

From Fig. 10-5, it is evident that at $\theta = \theta_1$, the voltage V_o is equal to $V' \cos \theta_1$. The detected voltage across the load is then equal to:

$$V_o = I_o R_L = \frac{V'R_L}{\pi r_d}(\sin \theta_1 - \theta_1 \cos \theta_1)$$

Substituting for V' yields:

$$V_o = \frac{V_{c_m}R_L}{\pi r_d}(\sin \theta_1 - \theta_1 \cos \theta_1) + \frac{V_{c_m}m_aR_L}{\pi r_d}(\sin \theta_1 - \theta_1 \cos \theta_1) \cos \rho t$$

The first term is independent of time and is a direct voltage with magnitude dependent on V_{c_m}. The second term varies with the modulating frequency and has an amplitude directly proportional to $m_aV_{c_m}$. This represents recovery of the original modulating signal without distortion.

The detection efficiency of a detector is defined as the ratio of the actual output voltage at the modulation frequency to the maximum possible value of the output. Note that to determine the maximum, assume that R_L is equal to infinity and capacitor C_L charges to the peak value of V_{c_m}. It is then obvious that the average V_o would equal V'. Then,

$$\eta_d = \frac{V_o}{V'} = \cos \theta_1$$

Substitution for V_o and V' yields:

$$\eta_d = \frac{\dfrac{V'R_L}{\pi r_d}(\sin \theta_1 - \theta_1 \cos \theta_1)}{V'}$$

and

$$\cos \theta_1 = \frac{R_L}{\pi r_d}(\sin \theta_1 - \theta_1 \cos \theta_1)$$

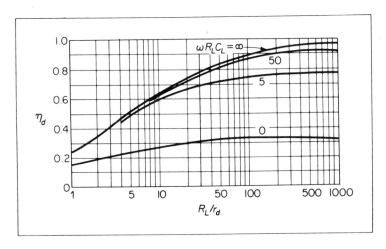

Figure 10-6 *Variation of Efficiency with C_L*

which simplifies to:

$$\frac{R_L}{r_d} = \frac{\pi}{\tan \theta_1 - \theta_1}$$

A family of curves shown in Fig. 10-6 can be obtained for the detection efficiency of a diode detector versus the ratio of R_L to r_d for assumed value of $\omega R_L C_L$.

INPUT IMPEDANCE OF LINEAR DETECTOR

The diode and load both appear as one resistive load to a tuned circuit input. The shunting effect of the diode and load tends to reduce the effective Q of the tuned circuit. The average input power to the diode and load is given by:

$$P_i = \frac{(V')^2}{2\pi r_d}(\theta_1 - \sin \theta_1 \cos \theta_1)$$

The input resistance of the diode and load is equal to the resistance that would dissipate the same input power with the same voltage applied. Consequently, the effective input resistance is given by:

$$R_{\text{eff}} = \frac{V'^2}{2P_i} = \frac{\pi r_d}{\theta_1 - \sin \theta_1 \cos \theta_1}$$

Let $\qquad \beta = \dfrac{\pi}{\theta_1 - \sin \theta_1 \cos \theta_1}$

and $\qquad R_{\text{eff}} = \beta r_d$

Using the relationship

$$r_d = \frac{R_L(\tan \theta_1 - \theta_1)}{\pi}$$

The ratio of the effective input impedance to the load impedance becomes:

$$\frac{R_{\text{eff}}}{R_L} = \frac{\tan \theta_1 - \theta_1}{\theta_1 - \sin \theta_1 \cos \theta_1}$$

The graphical relationships inherent in detector theory are shown in Fig. 10-7.

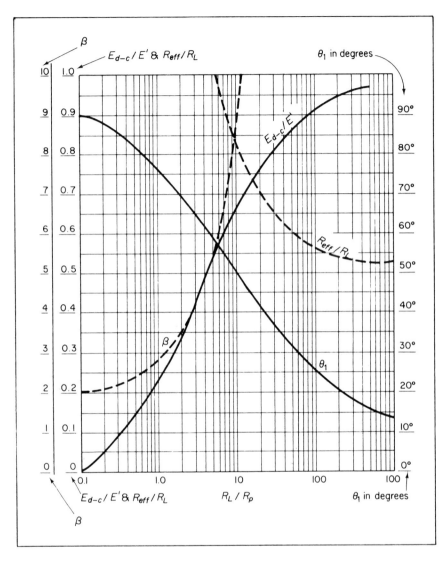

Figure 10-7   *Universal Relation in High-Level Detectors*

DISTORTION IN LINEAR DETECTORS

The $R_L C_L$ time constant of the diode load determines the rapidity with which the voltage across the load can change. If the effective diode resistance is small compared to R_L, the output voltage will tend to follow the modulation envelope.

For large C_L, the time constant of the discharge may be so large that the detector may not have time to discharge through R_L with the result that diagonal clipping will occur as shown in Fig. 10-8.

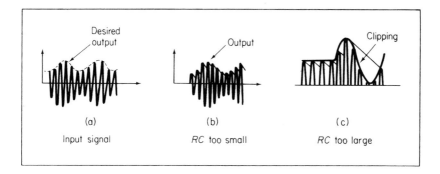

Figure 10-8 *Waveforms of Input and Output Voltages (a) Desired Output (b) RC Too Small (c) RC Too Large, Diagonal Clipping*

Analysis of the problem shows that the modulation envelope will be followed when the value of the load time constant is limited to:

$$\rho R_L C_L \leqq \sqrt{\frac{1 - m_a^2}{m_a^2}}$$

The selection of the circuit parameters is determined by the maximum modulating frequency and the percentage modulation. A second form of distortion—peak clipping—may occur when the coupling circuit is incorporated into the detector. Consider the circuit shown in Fig. 10-9.

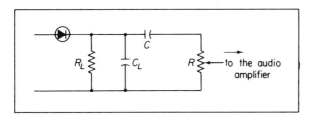

Figure 10-9 *Diode Detector Circuit*

The purpose of the coupling circuit is to permit the audio information to pass to the next stage, whereas the dc component is blocked by C. The dc component that is developed across the diode is ordinarily used for automatic gain control. The tap on R is variable so that the output of the detector may be controlled as desired as the audio voltage goes on to the next stage.

As a result of this complex network, the dc and ac circuit impedances are different, the ac impedance being smaller than that of the dc circuit.

The peak ac voltage, V_{ac}, is equal to:

$$V_{ac} = I_{ac}(\text{peak})R_{ac}$$

If clipping is to be prevented, the peak value of the ac current must equal I_{dc} for worst case conditions. Then,

$$\frac{\text{peak } V_{ac}}{V_{dc}} = m_a = \frac{R_{ac}}{R_L}$$

Note that R_{ac} is equal to:

$$R_{ac} = \frac{R_L R}{R_L + R}$$

The actual choice of the parameters is a compromise between detection efficiency and distortion. With a large input signal and a large R_{ac} to linearize the diode characteristic, it is possible to keep the distortion to one or two percent in envelope detection.

An illustrative problem will demonstrate the theory.

sample problem

In the circuit shown, determine the efficiency of detection, the effective load presented to the IF amplifier and the AVC voltage available at the output.

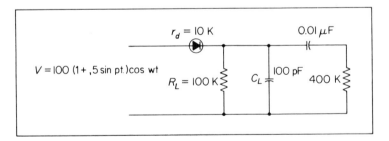

Solution:

> *Step 1:* The value of detection efficiency under dc conditions is found by the ratio of R_L to r_d equal to 10. From Fig. 10-7, the value is:
>
> $$\eta_{dc} = 65\%$$

The ac efficiency is found by the ratio of R_{ac} to r_d equal to 8. From Fig. 10-7, the value is:

$$\eta_{ac} = 60.5\%$$

Step 2: The effective load presented to the IF amplifier is from Fig. 10-7:

$$\frac{R_{ac}}{r_d} = 8 \qquad R_{eff} = .88(100)10^3$$

Step 3: To determine the dc or AVC voltage,

$$V_{dc} = .65 V' = .65\ (100) = 65\ \mathrm{V}$$

TROUBLESHOOTING ANALYSIS

There are essentially three major possibilities for diode detector failure. The first of these is the no-output condition. Generally, a no-output condition is limited to an open or short circuited component or diode. If the input transformer were open, stray capacitive coupling may feed sufficient signal through to produce an output. With an open detector load resistor or shorted bypass capacitor, the output is zero. When the circuit has no output, a resistance analysis (check) will rapidly locate the malfunctioning element.

The second possibility is the low-output condition. This condition can be caused by:

1. Low input signal
2. Faulty solder connections
3. Open input transformer
4. Open load bypass capacitor

The detector input and output are checked with a high impedance voltmeter. Since the stage does not amplify and the circuit efficiency is high, a dc output indicator of less than ten percent of the input indication would be symbolic of a detector problem.

The third possibility is that of output distortion. Since diode detection is assumed linear, a distorted output usually indicates circuit element changes. Amplitude distortion indicates nonlinearity in the detector. Distortion at high volume levels with a strong modulated signal would indicate peak clipping effects. *Fringe howl* or a tendency to oscillation would indicate poor radio frequency or intermediate radio frequency filtering. An examination of the waveform by an oscilloscope should locate the trouble rapidly.

AUTOMATIC GAIN CONTROL

In radio reception, the average amplitude of the modulated carrier varies. To provide a nominally constant output power, it is desirable to provide some method within the receiver to maintain the amplitude at the detector at some predefined level. This is done with automatic gain control (AGC) or automatic volume control (AVC) circuits.

The dc voltage across the output terminals of the diode detector is a measure of the signal intensity and is independent of m_a. The value of the dc voltage is:

$$V_o = \frac{V_{c_m} R_L}{\pi r_d} (\sin \theta_1 - \theta_1 \cos \theta_1)$$

The essentials of a circuit that provides AGC output are shown in Fig. 10-10.

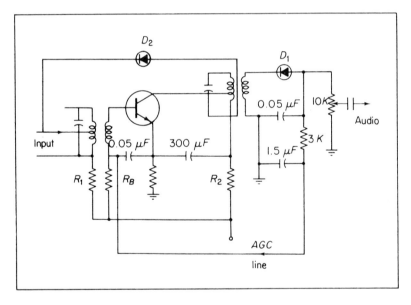

Figure 10-10 *A Simple AGC System*

A low-pass RC filter defined by the 3K resistor and the 15 μF capacitor is used to separate the dc voltage from the modulated wave. The dc voltage is then used as a control bias to vary the gain of the amplifiers before the detector.

In transistor amplifiers, several currents might be varied to produce the desired change in amplification. The most desirable parameter to vary appears to be I_e.

The diode D_2 provides additional control action when the amplitude of the input signal is large. At normal signal inputs, the diode D_2 is reverse biased by the voltage levels across R_1 and R_2. Typical operation of the AGC system decreases the collector current of the transistor Q and decreases the drop across R_2. At some preset level, D_2 will start to conduct. The insertion of a low diode resistance across the input tuned circuit reduces the signal amplitude. This method is quite effective as a backup method to the regular AGC system.

FREQUENCY MODULATED RECEIVERS

The transmission of an FM signal requires special circuitry to recover the audio information sent. Note that the original information has varying amplitudes that are incorporated into the FM signal. Consequently, the FM detector must be capable of translating these frequency variations back into an amplitude variation. In addition, the FM detector must be insensitive to amplitude variations. A method of limiting or reducing the signal to a constant amplitude utilizes a circuit called a *limiter*.

A gate limiting and drain saturation circuit is shown in Fig. 10-11.

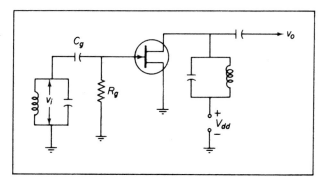

Figure 10-11 *Limiter Circuit*

The circuit is operated at zero gate voltage. The application of a signal input causes a dc voltage to appear across R_g. Under these conditions, the positive peaks of the signal are held clamped or constant at near zero voltage. If the drain electrode is operated at a relatively low potential, drain current saturation occurs. The FET limiter performance is shown in Fig. 10-12.

Another limiting circuit uses the emitter coupled limiter circuit shown in Fig. 10-13.

Assume Q_1 is nonconducting and Q_2 is conducting. The output voltage is equal to the saturation voltage of Q_2. This condition is maintained until signal e_i is applied increasing in a positive direction. Q_1 starts to conduct and causes the current through the emitter to increase with a resultant increase

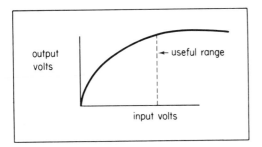

Figure 10-12 *Limiter Performance*

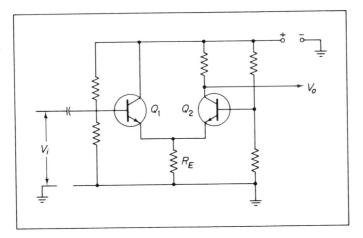

Figure 10-13 *Emitter Coupled Limiter*

in emitter voltage. Consequently, for a positive signal input with a peak-to-peak value equal to 2 V_q, the output voltage is limited to a value equal to V_{cc} (cutoff) and a negative value equal to saturation or zero. This action is demonstrated in Fig. 10-14.

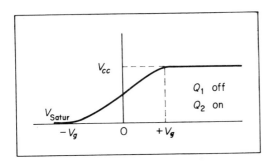

Figure 10-14 *Waveform in Emitter Coupled Limiter*

THE FREQUENCY DISCRIMINATOR

Frequency variations must be converted into amplitude variations. This is the first step in FM demodulation. A circuit for accomplishing this conversion is called a *discriminator*.

The simplest form of a discriminator is the slope detector, the mode of operation for which is shown in Fig. 10-15. In this case, the circuit is tuned to f_o rather than f_r.

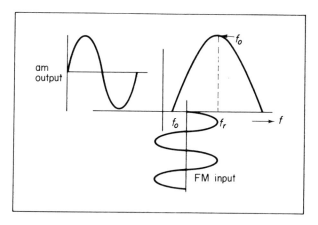

Figure 10-15 *Slope Detector*

As the carrier frequency fluctuates, the current in the detuned circuit varies accordingly, increasing as the input FM frequency approaches f_r and decreasing as the input frequency drops below f_o. The output voltage is an amplitude modulated signal somewhat distorted. Consequently, since this circuit produces a distorted output waveform, it is seldom used.

The distortion can be removed by using the discriminator that employs a double tuned transformer-coupled system as shown in Fig. 10-16.

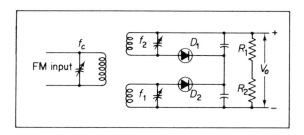

Figure 10-16 *Discriminator Using a Double Tuned Transformer*

The input tuned circuit is adjusted for carrier frequency f_o. The tuned circuit in series with D_1 is tuned to f_2 or somewhat higher than f_c, whereas the other tuned circuit is adjusted for f_1 or a frequency lower than f_c. When properly tuned, the output waveform is linear, as shown in Fig. 10-17.

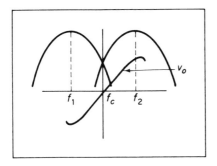

Figure 10-17 *AM Output from Discriminator*

The diode detectors give voltages proportional to each AM signal. The output voltage is equal to the difference of these two voltages.

The Foster Seely Circuit shown in Fig. 10-18 is a typical discriminator circuit. The primary and secondary circuits are both tuned to the same center frequency. If the primary and secondary loaded Q's are high (ten or greater) then:

$$\omega_o L_1 = \frac{1}{\omega_o C_1}$$

and

$$\omega_o L_2 = \frac{1}{\omega_o C_2}$$

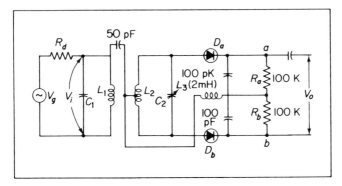

Figure 10-18 *Foster Seely Frequency Discriminator*

The internal resistance of the primary is neglected and the resistance of the diodes r_d is incorporated into the secondary. The equivalent circuit of the double tuned transformer-coupled circuit is shown in Fig. 10-19.

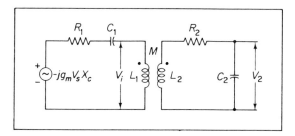

Figure 10-19 *Equivalent Circuit for Transformer*

The voltage V_i is evaluated by the equation:

$$V_i = \frac{-jg_m V_s X_{c_1}(jX_{L_1})}{R_1 + j(X_{L_1} - X_{c_1})}$$

Simplifying yields[1]

$$V_i = g_m V_s Q \omega_0 L_1 \left[\frac{1 + j2\delta Q - jk^2 Q}{(1 + j2\delta Q)^2 + k^2 Q^2}\right]$$

The term $jk^2 Q$ can be neglected with the resultant simplified equation given as:

$$V_i = g_m V_s Q \omega_0 L_1 \left[\frac{1 + jx}{(1 + jx)^2 + \gamma^2}\right]$$

where

$$x = 2\delta Q$$
$$\gamma = kQ$$

The secondary voltage of the double tuned circuit is given by:

$$V_2 = \frac{jg_m Q \omega_0 \sqrt{L_1 L_2} V_s}{(1 + jx)^2 + \gamma^2}$$

The secondary circuit operates through two diodes. The diodes D_a and D_b have applied voltages V_a and V_b with respect to ground. Thus,

$$V_a = V_i + \frac{V_2}{2}$$

$$V_b = V_i - \frac{V_2}{2}$$

[1]All definitions previously given in the section on tuned voltage amplifiers.

At the center frequency of the tuned circuit (f_o) δ is zero and the primary and secondary voltages are:

$$V_1 = \frac{g_m V_s Q \omega_o L_1}{1 + \gamma^2}$$

$$V_2 = \frac{j g_m V_s Q \omega_o \sqrt{L_1 L_2}}{1 + \gamma^2}$$

It is evident that V_2 leads V_i by 90 degrees. These relationships can be shown by the phasor diagram shown in Fig. 10-20.

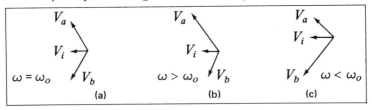

Figure 10-20 *Phasor Diagrams of Discriminator*

At the center frequency the voltage across the diodes is equal. At frequencies above resonance the phasor diagram becomes as shown in Fig. 10-20(b), whereas at frequencies below resonance, the resultant phasor diagram is shown in Fig. 10-20(c).

The diodes are connected to provide an output V_o that is directly proportional to the difference between $|V_a|$ and $|V_b|$. At resonance this difference is zero. The diode voltages are, respectively:

$$V_a = g_m Q \omega_o L_1 V_s \left[\frac{1 + jx + j\frac{\gamma}{2}\sqrt{\frac{L_2}{L_1}}}{(1 + jx)^2 + \gamma^2} \right]$$

$$V_b = g_m Q \omega_o L_1 V_s \left[\frac{1 + jx - j\frac{\gamma}{2}\sqrt{\frac{L_2}{L_1}}}{(1 + jx)^2 + \gamma^2} \right]$$

The difference between the two magnitudes is given by:

$$|V_a| - |V_b| = \frac{g_m Q \omega_o L_1 V_s}{\sqrt{(1 + \gamma^2 - x^2)^2 + 4x^2}} \left(\sqrt{1 + \left(x + \frac{\gamma}{2}\sqrt{\frac{L_2}{L_1}}\right)^2} \right. \\ \left. - \sqrt{1 + \left(x - \frac{\gamma}{2}\sqrt{\frac{L_2}{L_1}}\right)^2} \right)$$

For the optimum selection of parameter values ($\gamma = 1$, $L_2 = L_1$) the voltage magnitude difference appears as shown in Fig. 10-21.

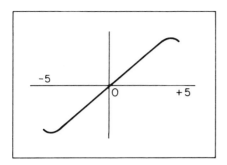

Figure 10-21 *The Discriminator Characteristic*

Note that any amplitude variation or modulation inherent in the signal V_s will reach the detector. Consequently, an effective limiter must precede the detector.

THE RATIO DETECTOR

The ratio detector is used extensively where noise free performance without a limiter is required. The essential portion of the circuit, adequate for explanatory purposes, is shown in Fig. 10-22.

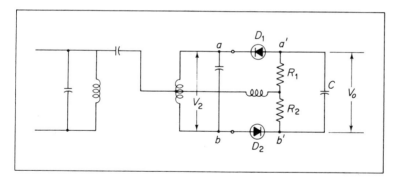

Figure 10-22 *The Ratio Detector*

It is evident that this circuit appears identical to the discriminator except that one diode is reversed. Assume the voltage across R_1 at f_o is equal to $-V$ volts and that the voltage across R_2 is also equal to $-V$ volts. Then the output voltage V_o is equal to $-2V$ volts. Now if the input signal is detuned slightly, then the voltage across R_1 is given by $-V + \delta V$ volts and the voltage across R_2 is given by $-V - \delta V$ volts. The net output voltage is still $-2V$ volts.

The voltages across R_1 and R_2 can be written respectively as:

$$V_1 = k_1 V_s$$
$$V_2 = k_2 V_s$$

For a given deviation, the output voltage is constant, but the ratio is given by:

$$\frac{V_1}{V_2} = \frac{k_1}{k_2}$$

and this ratio is independent of the signal amplitude.

A practical form of the ratio detector is shown in Fig. 10-23. At center frequency, the output voltage across R_1 is zero since the voltages across C_1 and C_2 are equal. When the frequency deviates off center, the voltages change across C_1 and C_2 and the voltage changes across R_1 at the deviation rate.

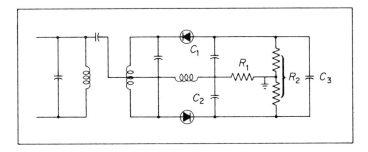

Figure 10-23   *Practical Ratio Detector*

However, for a long time constant R_2C_3, the output voltage across C_3 will not respond to rapid changes in input signal amplitude. This circuit is somewhat critical of adjustment for linear operation.

THE RADIO RECEIVER

The various circuits comprised by the radio receiver have been discussed and analyzed. The circuit of a commercial radio receiver is shown in Fig. 10-24. This receiver is designed to have sufficient sound for outdoor use, long battery life, high gain for good sensitivity and a minimum number of parts for reliability and economy.

To provide both high sound output and long battery life, a Class B push-pull output stage is used. The audio driver is transformer coupled to the output stage. Base bias is derived from the network R_{13} and R_{14}. R_{12} serves as a volume control. The zero signal battery drain is 8 mA. The maximum power output of this particular push-pull stage is 200 mW. All of the radio transformers should be wound on laminations of $1\frac{5}{8}''$ by $1\frac{3}{8}''$ and $\frac{1}{2}''$ stack size having an electrical efficiency of 80 percent. Smaller or less efficient transformers will degrade the electrical fidelity of the circuits.

A diode is used as both a detector and an AVC source. At low signal levels the bias of the first i-f stage depends on the voltage at the top or anode of CR 2. At strong signals, the bias depends on the voltage at the anode of CR 2 as well as the rectified voltage across CR 2. Thus, the voltage amplification of the i-f amplifier depends upon the signal level and the position of the volume control. The i-f amplifier consists of two stages. Signal is applied to the input of the transistors by connecting the secondaries of the transformers directly to the bases and through the capacitors C_5 and C_7 to the emitters. As far as the signal is concerned, this input puts the emitter resistors in the output circuit. In most cases, the emitter resistors are small when compared with the output load impedance and can be left unbypassed with very little noticeable loss in gain. For any given collector current, the value of the emitter resistors R_6 and R_{10} is a compromise between temperature stability and interchangeability on the one hand and the loss of collector voltage and gain on the other hand.

The frequency converter is a self-oscillating mixer that uses emitter injection. To supplement the AVC at strong signals, an overload diode circuit is used between the converter and the first i-f stage. The overload diode circuit provides approximately 50 percent of the gain control of the receiver.

The alignment and testing of an AM receiver requires at least three pieces of test equipment. They are:

1. An audio generator.
2. An r-f generator that covers the r-f and i-f frequencies of the receiver under tests and is capable of being modulated with audio frequencies.
3. An ac voltmeter or oscilloscope to measure the output signal. The oscilloscope can observe the output waveform.

The audio and r-f generators provide test signals for the receiver. The voltmeter and oscilloscope indicate the effect of these signals. The signals are fed into the receiver through coupling capacitors. This insures that the dc operation of the receiver is not affected by the insertion of test signals.

The value of the coupling capacitor used to couple the audio oscillator should be about 0.5 μF or larger. The capacitor used to couple the r-f generator should be about 0.01 μF or larger. In both cases, nonpolarized capacitors must be used and the voltage rating of these capacitors must be at least as high as the power supply voltage rating of the receiver.

Figure 10-24 *Six-Transistor Six-Volt Broadcast Receiver*

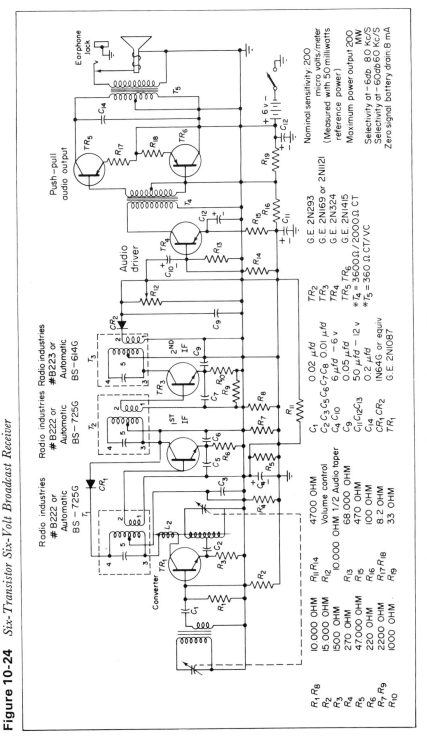

Courtesy of General Electric Semiconductor Products Department, Syracuse, N.Y.

The output signal voltage is generally taken across the loudspeaker and, therefore, the voltmeter is connected to this point. The test signals are adjusted to provide a measurable output signal. The alignment procedure is specified by a step-by-step procedure. Thus,

Step 1: Connect the r-f generator to the high side of the input tuned circuit using a 0.01 μF capacitor.

Step 2: Set the modulation of the r-f generator to 30 percent at a frequency of 1000 Hz. Set the r-f generator to the i-f frequency (455 kHz) and adjust the signal to produce an undistorted output signal.

Step 3: Set the volume control at maximum. Set the receiver tuning to a quiet spot in the band and adjust the i-f transformers for maximum output. Start with the last transformer first and work toward the converter.

Step 4: Connect the r-f generator to a short piece of wire and place the wire near the antenna.

Step 5: Set the generator frequency to the highest frequency in the AM band (1650 kHz). Adjust the signal to produce an undistorted output signal.

Step 6: Set the receiver tuning to the extreme high end of the AM band. Adjust the oscillator trimmer capacitor for maximum output signal.

Step 7: Reduce the generator frequency approximately 20 percent from the high end of the AM band. Set the receiver tuning accordingly and adjust the trimmer capacitors on all the r-f circuits for maximum output.

Step 8: Set the generator frequency and the receiver tuning to about 20 percent from the low end of the dial (AM band). Adjust the oscillator coil for maximum output while rocking the tuning capacitor back and forth through resonance.

The procedure is repetitive and the receiver is adjusted and readjusted until the tuning is completed.

problems

1. A square law detector has the following relationship:

   $$i = 2 \times 10^{-6}v^2 \text{ A}$$

   If $v = 5 \sin 2\pi \times 10^7 t + 2 \cos 2\pi \times 10^3 t$ V, tabulate the amplitudes of the frequency components.

2. A square law detector has the following relationship:

   $$i = 2 \times 10^{-3}v^2 \text{ A}$$

   If $v = 4 + 2 \sin 2\pi \times 10^6 t + .4 \sin 2\pi \times 2 \times 10^3\ t$ V, tabulate the amplitudes of the frequency components.

3. A square law detector has the following relationship:

   $$i = 4 \times 10^{-6}v^2 \text{ A}$$

   If $v = 20(1 + .3 \cos 2\pi \times 10^3 t) \cos 2\pi \times 10^7 t$ V, tabulate the amplitudes of the frequency components.

4. A square law detector has the following relationship:

   $$i = 2 \times 10^{-3}(2 + v^2) \text{ A}$$

   If $v = 3(1 + .75 \cos 2\pi \times 10^3 t) \cos 2\pi(10^6)t$ V, tabulate the amplitudes of the frequency components.

5. A square law detector has the following relationship:

   $$i = (4 + 2v^2)10^{-3} \text{ A}$$

   If $v = 10 (1 + .05 \cos 2\pi \times 10^4 t) \cos 2\pi \times 10^7\ t$ V, tabulate the amplitudes of the frequency components.

6. The voltage applied to a linear detector shown is given by:

   $$v_i = 20(1 + .4 \sin pt) \cos \omega t$$

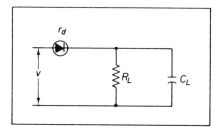

Find: (a) the detection efficiency $r_d = 5\,\text{k}\Omega$, $R_L = 100\,\text{k}\Omega$, $C_L = 100\,\text{pF}$.

(b) the effective resistance of the diode detector and load.

(c) the AVC voltage.

7. The input voltage to a linear detector is:

$$v_i = 25(1 + .6 \sin pt) \cos \omega t \text{ V}$$

$$r_d = 5\,\text{k}\Omega, \ C_L = 100\,\text{pF}, \ R_L = 200\,\text{k}\Omega$$

Find: (a) η_d (b) R_{eff} (c) AVC voltage.

8. In the circuit shown, the given data are:

$$v = 20(1 + .4 \sin pt) \cos \omega t \text{ V}.$$

$$r_d = 10\,\text{k}\Omega, \ R_L = 100\,\text{k}\Omega, \ R = 500\,\text{k}\Omega, \ C_L = 100\,\text{pF}, \ C = .01\,\mu\text{F}$$

Find: (a) η_d (b) R_{eff} (c) the AVC voltage.

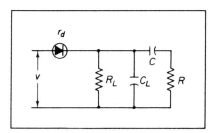

9. The applied voltage is:

$$v = 25(1 + .6 \sin pt) \cos \omega t$$

$$r_d = 5\,\text{k}\Omega, \ R_L = 200\,\text{k}\Omega, \ C = .65\mu\text{F}, \ C_L = 100\,\text{pF}, \ R = 250\,\text{k}\Omega$$

Find: (a) η_d (b) R_{eff} (c) AVC voltage.

10. The applied voltage is:

$$v = 50\,(1 + .3 \sin pt) \cos \omega t \text{ V}$$

$$r_d = 20\,\text{k}\Omega, \ R_L = 200\,\text{k}\Omega, \ C = .025\,\mu\text{F}, \ C_L = 150\,\text{pF}, \ R = 400\,\text{k}\Omega$$

Find: (a) η_d (b) R_{eff} (c) the AVC voltage.

11. Given the circuit shown, determine V_o for each part.

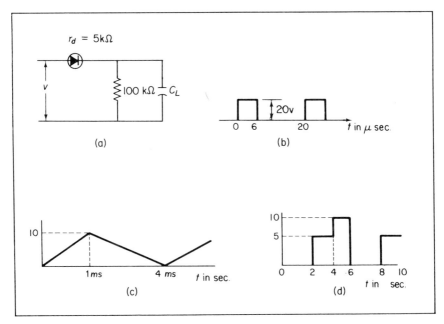

(a)
(b)
(c)
(d) $v = 10^5 (20)t \Big|_0^{5\ \mu\text{sec}}$

(e) $v = 10^{12}\ t^2 \Big|_0^{1\ \mu\text{sec}}$

12. A discriminator circuit has the given data:

$\gamma = 2$ $L_2 = \frac{3}{2} L_1$ $Q = 50$ $V_s = 2\ \text{V}$
$f_o = 10\ \text{MHz}$ $g_m = 3000\ \mu\text{mhos}$ $M = .05\ \mu\text{H}$

Find: V_o at 10.2 MHz.

eleven

RECEIVER CIRCUITS

INTRODUCTION

The transmitted signal received by the antenna of a radio receiver has an amplitude of a few microvolts. The signal energy delivered to a loudspeaker may be several volts or watts. The ratio of the amplitude of the output signal at the loudspeaker with respect to the amplitude of the signal at the antenna can be several million to a billion times.

Consequently, several stages of amplification are necessary and a desirable feature of these stages is that they produce zero distortion. The audio information transmitted has a frequency range of 20 Hz to 20 kHz and ideally should be reproduced exactly as transmitted. For the typical radio receiver, this *fidelity of reception* is not obtainable. Most receivers are able to receive and reproduce audio signals from 50 Hz to 5 kHz. More expensive AM (amplitude modulated) receivers are capable of reproducing signals in the range of 20 Hz to 10 kHz. For frequency modulated (FM) receivers, the audio frequency amplifiers should be capable of reproducing signals from 20 Hz to 15 kHz.

BASIC RECEIVER CIRCUITS

One of the simplest radio receivers manufactured is shown in Fig. 11-1. Note that it is simple in that the elements used are (1) a tuned circuit, symbolized by L and C (2) a diode detector, symbolized by D and (3) an audio amplifier using a PNP type transistor.

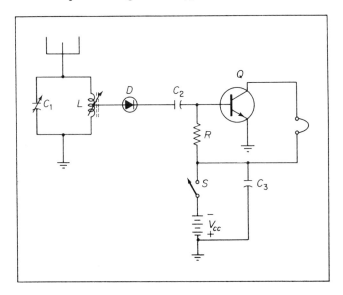

Figure 11-1 *Elementary Radio Receiver*

The tuned circuit is used to select the desired signal frequency. Since the coil or inductor is tapped, the Q of the tuned circuit is relatively constant and matches the effective impedance of the diode detector circuit. After detection by the semiconductor diode D, the demodulated signal is fed to the base of the audio amplifier stage. Resistor R performs a dual function: first, it is acting as load for the diode detector and, second, it is the base bias resistor for the transistor.

The operating power is supplied by the source battery symbolized by V_{cc} and is controlled by switch S for on-off operation. The amplified audio information is used to actuate the magnetic field of the head phones to produce an acoustic or sound output.

It is evident, therefore, that a radio receiver is a device used to reproduce audio signal information from a transmitted modulated radio frequency. The performance of this task can be specified by the analysis of its characteristics. These are sensitivity, selectivity, and fidelity.

Sensitivity: The sensitivity of a radio receiver is a measure of its ability to reproduce weak broadcast signals with satisfactory output strength. It is generally expressed as the amount of microvolt signal input to the antenna to produce a specified output either to the detector or to the speaker. In general, the greater the number of tuned amplifier stages, the better the sensitivity of the receiver. Thus, a receiver that requires 1000 microvolts input to the antenna to produce a one-volt input to the detector has a poorer sensitivity than a receiver that requires 250 microvolts input to the antenna for the same output. Typical sensitivity values will range from 250 μV to 5000 μV for receivers operating in the amplitude modulated broadcasting band.

Selectivity: The selectivity of a radio receiver is the ability of the receiver to separate a desired frequency from other signals. Selectivity is achieved by using high Q tuned circuits. In general, the greater the number of tuned circuits, the greater is the selectivity. The frequency response characteristic of the tuned radio frequency amplifier is also the selectivity curve. The sharper the response curve, the greater is the system selectivity. If the selectivity response characteristic becomes too sharp, the sidebands transmitted with the carrier modulated waveform may be clipped.

Fidelity: The fidelity of a radio receiver is the ability with which the system or portion of the system can faithfully reproduce the transmitted signal at the receiver output, usually the speaker.

ESSENTIALS OF RADIO RECEIVERS

The purpose of a radio receiver is to receive the transmitter broadcast radio signals for the recovery of the signal intelligence. The superheterodyne system is the basic circuit used in all modern receivers because of its higher gain and greater selectivity and sensitivity.

A superheterodyne system is one in which carrier frequency changes occur before the detection stage. A superheterodyne receiver is one that changes the frequency of the incoming carrier signal to a lower frequency called an intermediate frequency. This is produced by mixing the carrier frequency with the receiver oscillator frequency. The process of combining the two separate frequencies to produce a new output frequency is the process of *heterodyning*. Usually, the process of heterodyning is used to produce an output frequency equal to the difference of the two input frequencies. For example, when a broadcast radio frequency signal of 1000 kHz is heterodyned with an oscillator frequency of 1455 kHz, the output frequencies produced are (a) the original carrier frequency of 1000 kHz (b) the oscillator frequency of 1455 kHz (c) the sum of the two frequencies or 2455 kHz and (d) the difference of the two frequencies or 455 kHz. The circuit output stage is tuned to the lower of the two output frequencies or 455 kHz, which is called the intermediate frequency.

The block diagram of the basic superheterodyne circuit is shown in Fig. 11-2

The radio frequency (RF) stage is tuned to the desired carrier frequency. In the example shown, the carrier frequency is 1000 kHz. The signal is

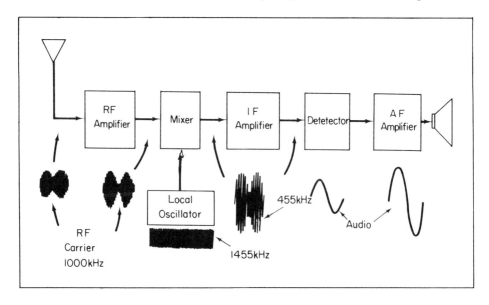

Figure 11-2 *Typical Superheterodyne Receiver Block Diagram*

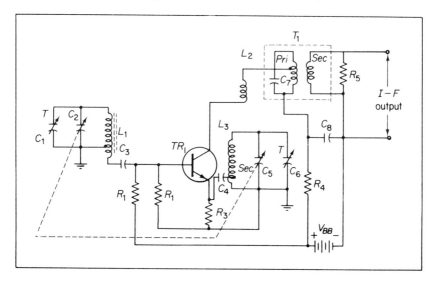

Figure 11-3 *Autodyne Converter Circuit Using an NPN Transistor*

amplified and coupled directly to the frequency converter or mixer. Also coupled into the mixer is the oscillator frequency and this frequency differs from the input signal by 455 kHz, thus providing an output frequency that is fixed at 455 kHz. The intermediate (IF) section of the receiver is always tuned to 455 kHz and is independent of the carrier frequency. Thus, as the tuned circuits of the RF amplifier are varied to select a different station, the oscillator circuit varies accordingly so that the mixer always produces an output of 455 kHz. Note that the tuning capacitors of the RF stage and the oscillator are ganged so that varying one capacitor also varies the other.

It is evident that the heterodyning process adds to the complexity of the receiver, but the advantages of this process are considerable. Thus, the receiver has higher sensitivity and selectivity plus uniform amplification over the broadcast band of the receiver.

Many transistor radio receivers use a single transistor to provide the functions of the mixer and oscillator circuits. This type of circuit is called the *autodyne frequency converter* and is shown in Fig. 11-3.

CIRCUIT THEORY

With the application of V_{cc}, current builds up inducing a voltage in the oscillator coil L_3. The frequency of oscillation is determined by the tuned circuit $L_3 C_5 C_6$. The oscillating signal frequency is then returned to the emitter of the transistor by means of C_4. Circuit elements $R_3 C_4$ provide emitter signal bias. L_1 is a ferrite core antenna that, in conjunction with C_1 and C_2, forms a selectively tuned circuit to provide an RF signal to the base of the transistor. The transistor is properly biased for linear operation. The incoming signal mixes with the oscillator signal producing a sum and difference of the two frequencies. The collector load comprises a primary and secondary tuned circuit that are tuned to the difference of the two frequencies or 455 kHz. This output frequency is maintained at 455 kHz since C_2 and C_5 are mechanically ganged together.

INTERMEDIATE FREQUENCY AMPLIFIERS

The mixer circuit consequently supplies an intermediate frequency (i.f.) to the amplifiers. This signal output from the mixer should be the desired RF signal converted to an i.f. signal. It is possible for the i.f. amplifier to amplify spurious or undesired signals. These spurious frequencies are those that are close to the i.f. frequencies of the receiver, harmonics of the desired RF signal and image signals. The effects of the spurious response may be readily ascertained by the listener. These effects may cause reception of two signals simultaneously, or whistles caused by the difference or beat of two signals, or reception of the same station at two different points on the dial.

The most troublesome of the spurious response is the image frequency signal. For example, assume that a receiver having an i.f. of 455 kHz is tuned to a station at 620 kHz. The oscillator frequency is set at 620 kHz plus 455 kHz or 1075 kHz. At the same time, an input frequency of 1530 kHz may be coupled into the mixer and also beat with the oscillator to produce an output signal of 455 kHz. It is obvious that the image frequency of any desired station is equal to the RF carrier plus twice the intermediate frequency. Thus, since the highest AM broadcast frequency is 1600 kHz, the image response can only occur at stations below 700 kHz.

The major function of an i.f. amplifier is to provide selective amplification. In general, the gain, bandwidth and selectivity requirements of the amplifier can be achieved with two i.f. amplifier stages. The two stages shown in Fig. 11-4 are essentially similar. The primary of each transformer is tuned and the secondary is untuned to match the collector output to the base input. Peaking of each i.f. stage is performed by adjusting the position of the iron core slug within the inductors.

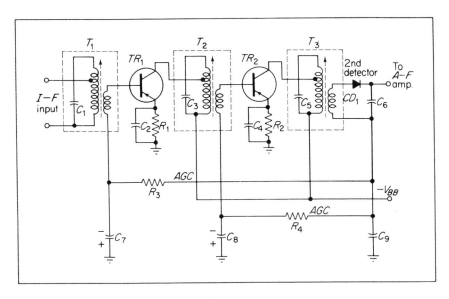

Figure 11-4   *Two-Stage i.f. Amplifier Circuit*

Neutralization of the i.f. stages is usually required since transistors operating at high frequencies are inherently unstable. In general, whether an i.f. stage will require neutralization depends on the size of the collector to base capacitance. In transistors specifically designed for i.f. operations, the internal collector to base capacitance is so low that neutralization is unnecessary.

AUTOMATIC VOLUME CONTROL (AVC)

To maintain a constant output level, the receiver would require continuous adjustment of the manual volume control each time the received signal varied in strength. The purpose of the AVC circuit sometimes called AGC (automatic gain control) is to provide a simple control of output sound volume from a speaker without blasting on strong stations or fading on weak stations. An AVC circuit is shown in Fig. 11-4, also.

AVC is achieved by rectifying the signal itself. The dc component of the signal is then directly proportional to the average modulated carrier amplitude. A portion of the dc voltage is then applied to the base input terminals of the RF i.f. transistors, changing the operating point of the circuit with a consequent change in the current amplification.

theory

When the incoming signal is strong, the dc component of the rectified signal is high thus decreasing the base voltage and reducing the emitter current and emitter voltage. It is evident that the stage amplification has decreased when the incoming signal is weak, the dc component fed back is low, thus increasing the base voltage and causing an increase in the emitter current and voltage. The time constant for the AVC circuit is specified by $R_3 C_7$ and $R_4 C_8$. Typical values for the time constant vary from 100 μsec to 1 msec.

AUDIO AMPLIFIERS

The function of the audio amplifier is to amplify the audio signals with fidelity. In the design and analysis section, the important considerations of this type amplifier were its linearity, frequency response and power output capabilities.

It is evident that the power output capabilities must be adequate to drive a loudspeaker or any other sound transducer. The power gain must be sufficient for the detector output to drive the audio amplifier stage to maximum output over the entire range of the audio frequency spectrum. The frequency response of the audio stage is affected by the frequency limitations inherent in the transistor itself and then by the reactive elements used for bias and coupling.

The tonal quality of the receiver depends on the linearity of the audio amplifier, the type of loudspeaker used and the linearity of the output transformer.

CLASS B OPERATION

The efficiency of Class B circuits is comparatively high and is commonly used for most portable applications. Class B operation requires the use of two transistors for high output and reduced distortion.

A brief review of the principles of Class B operation will now be presented. Refer to Fig. 11-5. Generally, a transistor operating in Class B draws current for one-half of the total cycle. The average current depends on the amplitude of the audio signal. When the signal is low or weak the current is small. This condition is desirable because the system is highly efficient.

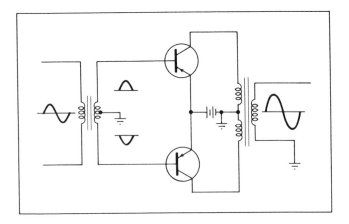

Figure 11-5 *Class B Amplifier*

With two transitors operating at Class B, the output currents should ideally be a sinusoid as shown. However, in practice the system nonlinearity produces a distorted output called *crossover distortion,* as shown in Fig. 11-6.

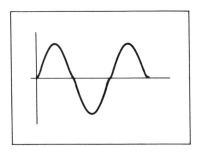

Figure 11-6 *Crossover Distortion*

To avoid crossover distortion, the transistors are biased to operate in the linear region of their characteristics. A complete circuit of a Class B operator using typical values is shown in Fig. 11-7.

The bias is applied from the voltage divider resistors R_1, R_2 and R_3. Resistor R_2 is a thermistor that is temperature sensitive and maintains the proper bias over the required temperature range of variation.

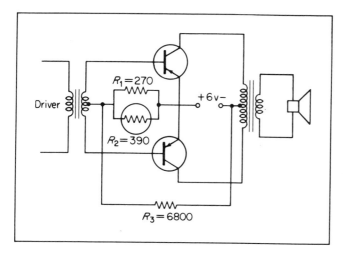

Figure 11-7 *Typical Class B Amplifier*

A method to eliminate the output transformer utilizes a Class B amplifier that is directly coupled to a tapped speaker load. This arrangement is shown in Fig. 11-8. The tap should be placed at the electrical zero of the speaker. The performance of this circuit is identical to the transformer-coupled circuit except that proper phasing of the output current is obtained by using the tapped speaker. Note that at any given time, only one-half of the speaker is used. Therefore, the resultant output sound power is approximately half of the sound power that results if the same electrical power were applied to the entire speaker. It is evident, therefore, that there is reduced output and efficiency in this circuit. This disadvantage is counterbalanced by several advantages. First, the cost and space requirements of the output

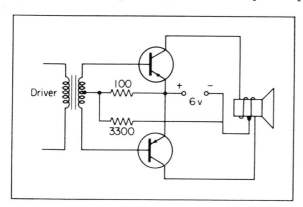

Figure 11-8 *Class B Amplifier with Tapped Speaker Coil*

transformer are eliminated. Second, the frequency limitation of the transformer disappears. Third, the impedance of the speaker is resistive and, consequently, provides sufficient system stability to eliminate an expensive thermistor.

TYPICAL SUPERHETERODYNE RECEIVERS

The superheterodyne circuit is the basic circuit used in portable transistor radio receivers. Fig. 11-9 shows the complete circuit of a radio receiver. Note that the system contains a 2N412 converter, two stages of i.f. amplification using 2N410 transistors, an overload diode designated by CR_1, a 1N60 second detector designated by CR_2, a 2N408 audio frequency driver stage and two 2N408s used as a Class B push-pull audio output stage. The transistors used are all PNP types and are connected in the grounded emitter circuit configuration except for the oscillator section of the converter, which operates as a grounded base circuit.

The tuning range is from 540 kHz to 1600 kHz with an i.f. stage tuned to 455 kHz. The receiver is capable of delivering a maximum output of 200 mW and an undistorted output of 150 mW. The receiver operates from a 4.5 V power source. The numbers above the stage indicate the amount of amplification each circuit provides.

converter circuit   The incoming radio frequency signal picked up by the antenna is coupled to the base of the converter transistor 2N412. There is no radio frequency preselector stage. The inductance of the antenna L_1 resonates with C_{T_1} for radio frequency tuning. The ac signal flowing through the primary of the oscillator coil causes current to flow through the secondary winding, which resonates with C_{T_2} (mechanically ganged to C_{T_1}). The oscillator frequency is coupled to the emitter by C_2 (.01 μF) resulting in non-linear operation of the transistor 2N412. The mixer section of the converter operates as a grounded emitter circuit with the radio frequency signal entering the base and the oscillator signal injected into the emitter. Resistor R_3 provides the operating bias and temperature stabilization for the circuit.

To supplement the AVC at strong signals, an overload diode circuit CR_1 is used between the converter and the first intermediate frequency amplifier stage.

intermediate frequency amplifiers (i.f. stages)   The difference frequency between the radio frequency and the oscillator frequency is selected by the intermediate frequency transformer. This receiver uses two stages of i.f. amplification having a total gain of 55 db or a signal amplification from 10 μV to 8 mV. Impedance matching between stages is obtained by tuning the primary winding, which is properly tapped. The secondary windings are connected directly to the bases and through C_3 and C_5 to the emitters.

Figure 11-9 *Superheterodyne Receiver*

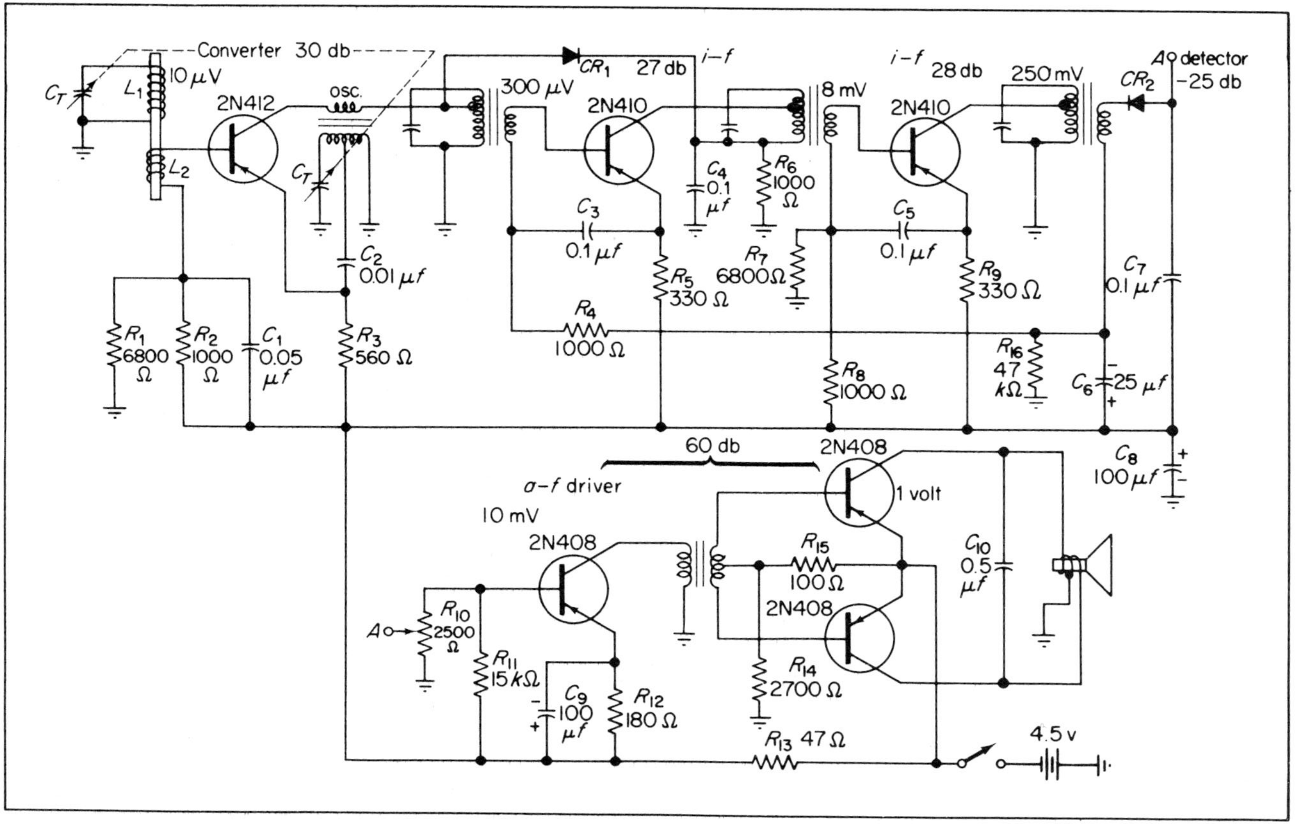

The emitter resistors R_5 and R_9 provide the required operating bias and stability for the i.f. amplifier stages. In this case, the emitter resistors are small compared to the load output impedance and can be unbypassed without reduction in amplification. When a strong signal is applied to the primary of the first i.f. amplifier, it causes the voltage across the base bias resistor to increase. This increase in voltage causes the overload diode to conduct with a consequent reduction in the stage amplification.

detector and volume control circuit   The audio frequency output of the diode detector is used to drive the first stage of audio frequency amplifiication. The dc output is used to provide automatic volume control (AVC). The load resistance for the diode detector is the volume control R_{10}. The amplitude of the output signal is controlled by varying a portion of the rectified signal voltage that is applied to the base of the first i.f. amplifier stage. With a weak signal applied to the antenna, the bias voltage on the first i.f. amplifier stage depends on the voltage across the volume control. With strong signals applied to the antenna, the bias voltage depends on the volume control voltage plus the rectified voltage across the overload diode CR_1. To maintain a proper time constant for transistor operation and to prevent the audio from being blocked, a large value of capacitance is required for the demodulating capacitance (C_7).

audio amplifier section   The audio frequency output from the detector is coupled directly to the base of the first audio frequency driver stage. The base bias is derived from the volume control R_{10} and resistor R_{11}.

The audio driver is transformer coupled to the audio output stages. The output transistors are operated in Class B push-pull to provide an undistorted output of 150 milliwatts. The proper forward bias is supplied for both output transistors by R_{14} and R_{15} respectively. The voice coil of the permanent magnet type speaker is center tapped to ground and eliminates the need for both an output transformer and a temperature sensitive bias network.

FREQUENCY MODULATED RECEIVERS (FM)

The broadcast amplitude modulated stations are operated over a band of frequencies assigned by the Federal Communications Commission (FCC) as 540 kHz to 1600 kHz. The broadcast frequency modulated (FM) stations operate over 88 MHz to 108 MHz. One of the major advantages of the FM system as compared to the AM system is the large reduction in the reception of undesired signals such as noise, static and so forth.

In AM reception, external noise affects the amplitude of the signal producing unwanted sounds in the output of the receiver. In some cases, the noise produced is so severe that the reception of the desired signal becomes impossible. On the other hand, FM reception is not affected by these external noise variations since they are leveled out by the limiter discriminator action. The receiver output is comparatively free from noises caused by static.

The FM receiver utilizes the same superheterodyne principles as the AM receiver, but the higher frequencies in the FM receiver create some additional circuit problems. Note that the FM receiver requires the additional stages of a limiter and a discriminator to perform properly.

A block diagram of an FM receiver is shown in Fig. 11-10.

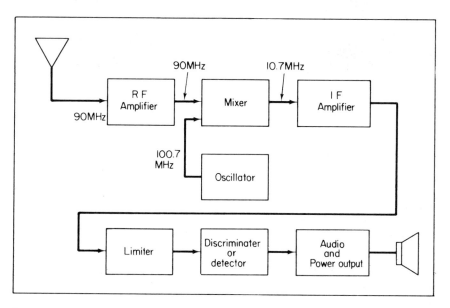

Figure 11-10   *Block Diagram of an FM Receiver*

Note that the FM signal is picked up by the antenna and fed to the RF stage. It is amplified and transferred to the mixer stage. Simultaneously, a signal developed by the oscillator is also fed into the mixer. The result of the combination within the mixer is shown in Fig. 11-11. The difference frequency is called the i.f. frequency and is amplified in the i.f. amplifier stage. The i.f. frequency for FM receivers is generally 10.7 MHz. The amplified i.f. signal is now passed on to a limiter where any amplitude variations are removed. The detector stage removes the carrier frequency and permits the audio signal to continue toward the speaker. The detector output signal is then fed to the conventional audio amplifiers and, finally, to the loudspeaker.

The function of the FM antenna is to provide the maximum possible signal voltage to the input of the RF amplifier. The antennas used in FM

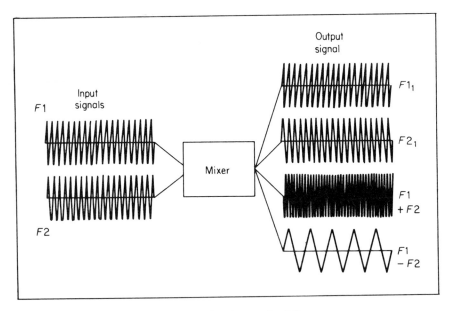

Figure 11-11   *Input and Output Waveforms of a Mixer*

receivers are cut to selected lengths to provide a proper match to the input RF amplifier. If the receiver is used in a strong signal area an antenna incorporated into the power cable will provide satisfactory results. In moderately strong signal areas the typical 300 ohm TV transmission line antenna can be used to provide satisfactory reception. For weak signal areas a commercial type of FM antenna is usually required for satisfactory reception. In all cases, the antenna possesses directional characteristics and requires experimental placement for good reception. If the input impedance of the RF amplifier is unequal to 300 ohms, the signal is fed or coupled to the input circuit through a matching transformer.

The radio frequency amplifier performs the same function in the FM receiver as it previously performed in the AM receiver; that is, to increase the *sensitivity* and *selectivity* of the receiver. The principal functions of the RF stages are to discriminate against undesired signals and to increase the amplitude of the weak signals.

frequency converters   Converters are essentially mixers that convert the combined RF and oscillator signals to an output i.f. signal. A typical converter circuit is shown in Fig. 11-12.

The RF signal input is applied to the antenna input terminals and through the matching transformer T_1 to the base of the transistor. The oscillator is a tickler feedback circuit having a tuned circuit (L_1C_2) with L_2 comprising a tickler feedback coil. Oscillation takes place as a result of the degree of coupling of the feedback signal from L_1 to L_2. Note that both L_1

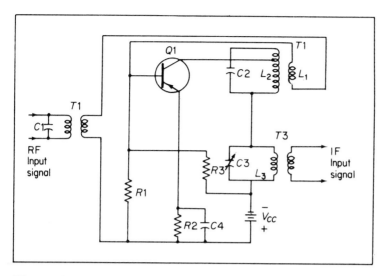

Figure 11-12   *Schematic Diagram of a Converter*

and L_2 are mutually coupled to L_3 so that the reflected impedance from L_3 controls the amount of mutual coupling that exists between L_1 and L_2. The reflected impedance in turn is determined by the resonant frequency of the tuned circuit $L_3 C_3$.

As a result of the oscillator feedback, both the input RF signal and the oscillator signal are applied in series between the base and emitter of the transistor. The transistor is biased for nonlinear operation and consequently produces an output equal to the difference of the two heterodyned frequencies. This difference frequency is selected by the tuned circuit in the collector $(L_3 C_3)$ and is inductively coupled to the i.f. stage.

intermediate frequency amplifiers (i.f.)   The i.f. amplifier stages used in an FM receiver are similar to that of the AM i.f. amplifier except that the FM circuit must have a higher frequency and a wider bandwidth. The standard i.f. amplifier frequency is 10.7 MHz as compared to the AM i.f. frequency of 455 kHz. The ideal and typical overall response for an i.f. section is shown in Fig. 11-13. The typical response characteristic illustrates the effect of varying the coefficient of coupling. It is evident that because of loading effects at the higher frequencies, the system amplification will be lower for an FM amplifier than for an AM amplifier circuit.

This system decreases in amplification dictates the usage of at least two or three stages of intermediate frequency (i.f.) amplification in an FM receiver.

Because of the higher operating frequency, the tuned circuit components (L and C) are comparatively small in value. I.f. transformers can be adjusted by varying either the capacitor or the inductor. In general, the

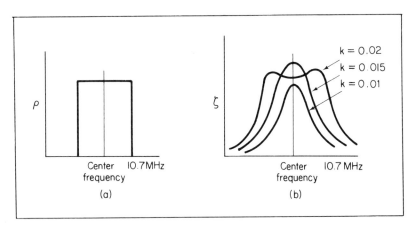

Figure 11-13   *(a) Ideal Frequency Response (b)   Typical Response*

transformer is tuned by varying the position of an iron slug within the inductor by means of an adjusting screw.

limiters   Limiter stages are used after the last i.f. stage in FM receivers to eliminate amplitude variations caused by fading or by noise in the waveform. A simple illustration is shown in Fig. 11-14.

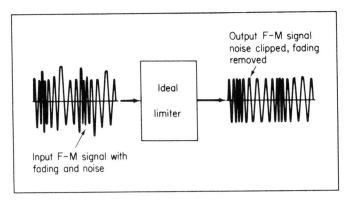

Figure 11-14   *Effect of Limiter Circuit on Waveform*

A somewhat better representation of the ability of the limiter to remove amplitude variations is shown by the curve in Fig. 11-15.

For all signal inputs to the antenna above a certain minimum voltage, the limiter produces a constant output voltage eliminating all variations. This region is shown to the right of point 2 in Fig. 11-15. In the region from 1 to 2, however, the stage is operating linearly.

Input voltages in this region will be amplified and will produce different output voltages. Thus, any signal that is too weak to drive the limiter

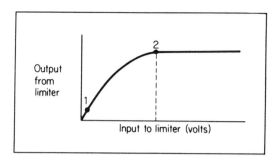

Figure 11-15   *Limiter Characteristic*

properly will result in amplitude variations with consequent distortion in the output. It is evident, therefore, that for proper operation the input signal to the antenna must be greatly amplified to provide adequate signal strength to produce a constant output.

A large reduction in amplitude variations can be obtained with a single stage limiter. Almost perfect limiter operation can be achieved by using two stages.

detectors   The FM detector converts the frequency variations of the FM signal to an audio output signal. In modern FM receivers, the *Foster Seeley* discriminator, sometimes called a *phase discriminator*, is one of the most widely used type of FM detector circuit and is shown in Fig. 11-16. Note that this type of circuit requires a center-tapped transformer.

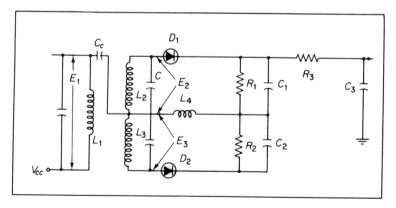

Figure 11-16   *Foster Seeley Discriminator*

theory

The output voltage V_1 of the limiter stage is also across L_4. The tuned circuit of the Foster Seeley discriminator (L_2, L_3 and C) is tuned to the i.f. frequency of the limiter or 10.7 MHz. Consequently, voltages V_2 and V_3

are exactly 90 degrees out of phase with V_1 (one is leading and the other is lagging). Note that V_2 is equal in magnitude to V_3, but 180 degrees out of phase with each other. Consequently, at the resonant frequency of 10.7 MHz, diodes D_1 and D_2 have equal voltages applied producing voltages equal and opposite in polarity voltages across R_1 and R_2 respectively. The net result of this action is to produce a zero output voltage at 10.7 MHz.

For frequencies above or below 10.7 MHz, the phase relationships between V_2 and V_3 with respect to V_1 shift from 90 degrees resulting in one diode receiving a higher voltage than the other diode. Under these conditions the voltages across R_1 and R_2 are unbalanced and produce an output whose value is dependent on the frequency variations. The mathematical analysis of this circuit has been performed in a previous chapter.

The time constant $R_3 C_3$ must be large enough to prevent any fluttering of the signal resulting from fast variations of the automatic frequency control bias circuits. Once the audio signal has been recovered, it is coupled to the volume control and audio amplifier stages for further amplification and output to the speaker.

ratio detector circuit   The basic circuit of a ratio detector is shown in Fig. 11-17. Note that this circuit is a modification of the Foster Seeley discriminator.

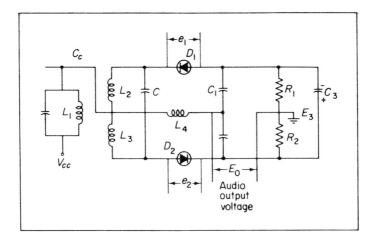

Figure 11-17   *Ratio Detector Circuit*

The ratio detector differs from the Foster Seeley discriminator in that: (1) diode D_1 is reversed and is now in series with diode D_2, (2) a large electrolytic capacitor is across the load resistors R_1 and R_2 and (3). The positive audio output voltage is taken at the junction of L_4, C_1 and C_2 with respect to ground.

The voltages across R_1 and R_2 are developed in exactly the same manner as described in the discriminator circuit; however, capacitor C_3 charges up to voltage V_3. When the peak to peak input voltage applied to the diodes is less than voltage V_3, the diodes become nonconducting and the signal output is zero. A load voltage appears only when the incoming peak to peak signal is large enough or larger than V_3. Consider that the input voltage to diode D_1 is e_1 and to D_2 is e_2.

The sum of the voltages across C_1 and C_2 must equal the voltage across C_3 for amplitude limiting conditions. When the incoming signal is at 10.7 MHz, the two voltages developed are equal and the output voltage is zero. Assume that v_1 is larger than v_2. The voltage across C_1 is greater than the voltage across C_2, producing a positive output voltage. When v_1 is less than v_2, an instantaneous negative output voltage is produced. The total output voltage is a constant and always equal to V_3, but a change in voltage across C_1 and C_2 must divide proportionally; thus the resultant name "ratio detector."

A dc component is present at the junction of diode D_1 and C_1 and can be used for automatic frequency control circuitry. The value of this dc component depends on the amount of difference between the input frequency and the 10.7 MHz frequency.

preemphasis and deemphasis When AM or FM waveforms are transmitted, noise and other extraneous disturbances are incorporated into the audio information. These undesired signals are generally more predominant at the higher audio frequency spectrum. Consequently, a circuit called a *preemphasis circuit* is inserted between the audio amplifiers and the FM transmitter to counteract the noise conditions. Emphasis is added to the higher frequencies in the transmitter to provide a better signal to noise ratio.

It is necessary, however, that the receiver use the inverse or a deemphasis network. This network must have an inverse characteristic to the preemphasis network so that both magnitude and frequency of the original signal are properly recovered. The two circuits must have the same time constant. According to standard FM broadcasting practices, the standard time constant for the preemphasis circuit is 75 μsec. The deemphasis circuit in the receiver usually employs the same time constant of 75 μsec.

audio frequency The audio frequency section of an FM receiver must have the same requirements as the AM receiver. Since the FM transmitter can broadcast all frequencies in the audio spectrum, the design requirements are a bit more exacting. All of these considerations have been thoroughly discussed and analyzed in previous chapters.

FM receiver The circuit diagram of an FM tuner is shown in Fig. 11-18. This particular circuit uses silicon NPN transistors for the RF stage to provide good sensitivity and selectivity. The transistor used for the stage is the 40242 connected as a common emitter circuit. The 40243 transistor is used as a mixer and is also operated as a common emitter circuit. The local oscillator generates a signal that is coupled into the base of the mixer. This oscillator signal voltage is provided by the 40244 transistor oscillator circuit.

The i.f. section comprises three stages that use either three 40245 or 40246 transistors in a common emitter circuit configuration to provide extremely high gain. The three double-tuned i.f. transformers T_1, T_2 and T_3 are adjusted for a bandwidth of 300 kHz.

The 1N295 diode and associated components in the collector circuit of the third i.f. amplifier develops a negative voltage that is proportional to the RF input signal. This voltage is used to trigger a noise immunity circuit. If desired, the negative voltage can also be used for automatic gain control by supplying this voltage to the base of the RF amplifier. This special connection should provide a wide AGC bandwidth for strong signals. Conse-

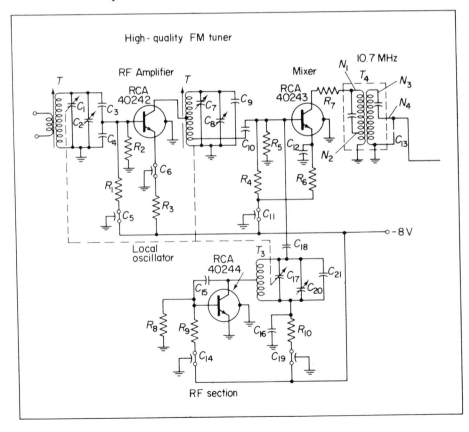

Figure 11-18 *FM Tuner (Courtesy RCA Semiconductor Products)*

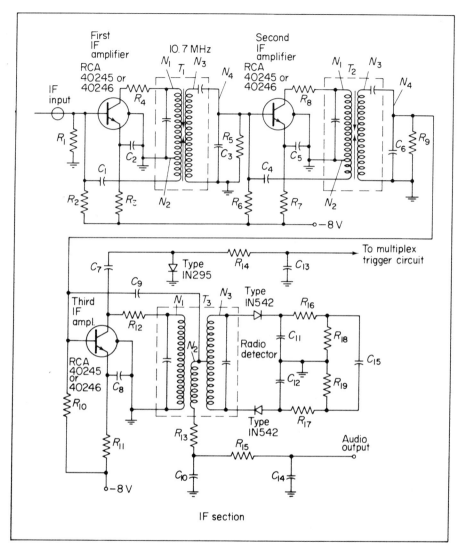

Figure 11-18 (Cont.)

quently, the third i.f. stage can completely limit the signal before appreciable AGC is developed.

FM detection is accomplished by the ratio detector circuit. This circuit has a matched pair of 1N542 diodes and associated components, R_{15} and C_{14} (3.9 kohms and .02 μF) in the detector output circuit form a standard FM deemphasis circuit of about 78 μsec.

A

INTEGRATED

CIRCUITS

INTRODUCTION

An integrated circuit is a system in which all circuit elements such as transistors, diodes, resistors and capacitors are contained within a single chip of silicon. Such circuits are produced by the same processes used to manufacture individual transistors and diodes. In the integrated circuit, various devices are isolated from each other by isolation diffusion within the crystal chip and are interconnected by an aluminum layer that serves as wires.

A method of batch processing called planar processing permits the manufacture of large numbers of ICs at low cost. The principal steps in integrated circuit technology are:

342

1. Crystal growth

2. Oxidation

3. Photolithography

4. Diffusion

5. Metallization

6. Encapsulation

The standard integrated circuit components are transistors, diodes, resistors and capacitors. The parameters of components in integrated circuits cannot be controlled as accurately as in discrete elements. However, identical components such as two transistors having the same h_{fe}'s are automatically produced.

EPITAXIAL

Various chemical processes are used to prepare chemically pure silicon from its unrefined state such as sand. A small crystal seed is inserted into the *silicon melt* and slowly rotated and withdrawn. The liquid silicon that is attached to the seed freezes as it is slowly withdrawn and a crystal is grown. The crystal is then cut into wafers 25 μm thick and 5 to 8 cm in diameter. After polishing and cleaning, a thin layer of silicon oxide is formed over the entire wafer.

A thin layer of silicon of a different conductivity type is grown onto the wafer. Such a layer is called an *epitaxial layer*. Silicon dioxide plays a major role in the fabrication and in the stability of silicon wafers since it prevents the diffusion of impurities through the wafer.

In Fig. A-1 the silicon wafer is shown with a silicon dioxide layer and photographic mask.

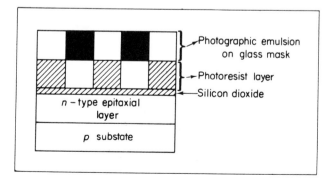

Figure A-1 *Exposure of Photoresist Through a Mask*

A photographic mask is placed over the coated wafer. An intense light source is applied through the mask causing the photoresist to become polymerized under the transparent regions of the stencil or mask. After exposure, the mask is removed and the photoresist is developed by dissolving the unexposed photoresist in a special solvent. The photoresist that is not removed in the development is now *fixed* or *cured*, and becomes insoluble to certain etching solutions.

The resultant chip is immersed in an etching solution that removes the oxide from the areas through which the impurities are to be inserted for diffusion. After etching and diffusion of impurities, the resist map is removed as shown in Fig. A-2, leaving the hardened photoresist.

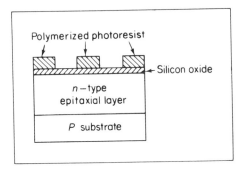

Figure A-2 *Photolithographic Process Resulting in Polymerized Photoresist*

The oxide layer now has open areas called windows that correspond to the open areas in the photoresist.

The manufacture of IC components requires the introduction of selective impurities in controlled amounts to be diffused into the silicon

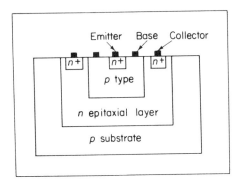

Figure A-3 *Integrated Transistor, npn Type*

crystal. A planar transistor is formed by first depositing an epitaxial *n*-type layer on a *p*-type substrate. The second step is to produce windows by means of the oxidation and photolithographic processes. The third step is to employ solid state *p*-type diffusion at the windows. Solid state *p*-type diffusion requires that the silicon crystal be heated to about 900° to 1200°C, so that impurity atoms can be injected and diffuse evenly throughout the crystal. The fourth step is to incorporate an *n*-type emitter diffusion. The resultant cross section is shown in Fig. A-3.

Note that the *n*-type collector is electrically separated from the *p*-type substrate by a junction forming a diode. This diode is normally reverse biased.

METALLIZATION

Evaporated aluminum is generally used to make connections on the semiconductor surfaces. The surface is prepared by growing an oxide and coating the surface. The oxide is removed using the photolithographic process where contact to the silicon surface is necessary.

Aluminum is then evaporated over the entire surface of the wafer and the excess is removed using the photoresist technique. Once the unwanted areas of aluminum are removed, the circuit is encapsulated to prevent damage to the chip from dampness and traditional mechanical hazards of handling and shipping.

Integrated circuits, in general, are constructed from NPN transistors, diodes and resistors. The reason these devices are used is that these components are the simplest and cheapest to fabricate. Note that the electrical characteristics of a transistor depend on the size and geometry of the device, the amount of doping and the basic silicon material.

It is desirable to minimize component size. Because of the nature of the photolithography process, it is not possible to make a photographic image smaller than a wavelength of light. This kind of optical limitation on image size is called *resolution limiting*.

A further practical limitation on minimum element spacing is imposed by the requirement that each photoresist mask align with other masks. The contact to a transistor base must not overlap into either the emitter or collector regions.

INTEGRATED RESISTORS

A resistor in an integrated circuit is often fabricated by utilizing the bulk resistivity of the diffused areas. The resistor is made during the transistor base diffusion or *p*-type base diffusion. The doping and thickness of the resistor are fixed by the base requirements. The resistance, therefore, is a function of the length and width. Thus,

$$R = R_s \frac{l}{A}$$

where

R_s = sheet resistivity, which depends on the thickness and resistivity of the region.

l = length of the diffused region.

w = width of the diffused region.

INTEGRATED DIODES

Diodes are common components in all IC systems for application in digital and pulse circuits. They can readily be obtained from transistor fabrication simply by shorting the collector to base to form an emitter–base diode or to use the emitter to base without shorting or maintaining the collector open circuited.

The choice of the diode type used depends on circuit application. Generally, the collector is shorted to the base, provided that the reverse breakdown voltage of the base to emitter is not exceeded in circuit applications.

INTEGRATED CIRCUIT CAPACITORS

Integrated circuit capacitors may readily be formed by using the junction capacitance of a reversed *p-n* junction. A cross-sectional view of a junction capacitance is shown in Fig. A-4.

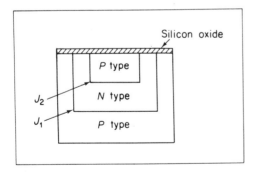

Figure A-4 *Diffused Junction Capacitor*

Whenever a semiconductor junction is formed a junction capacitance results. The capacitor is formed by reverse biasing the *p*-type and *n*-type regions, labeled J_2 on Fig. A-4. An additional capacitor is formed by junction J_1. The capacitor value of such a structure is given by:

$$C = 0.885 \frac{KA}{d} \text{ pF}$$

where:

      A = junction area in square centimeters

      K = dielectric constant of insulating material

      d = depletion layer thickness in cm

    The dielectric constant for silicon is 12. Note that capacitance C is a function of the junction area and inversely proportional to the thickness of the depletion region. The voltage input must have the negative terminal applied to the substrate, since the junction J_2 must be reverse biased for proper operation.

    A side view of a simple monolithic integrated circuit and the resultant equivalent circuit is shown in Fig. A-5.

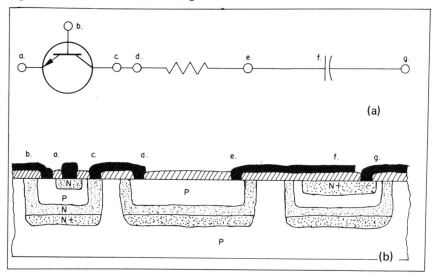

Figure A-5   *(a) Integrated Circuit (b) Equivalent Circuit
(Courtesy of RCA Service Co.)*

    The method used to construct this integrated circuit or chip was to start with a p-type crystal wafer upon which was epitaxially grown an $n+$ layer and an n layer. The basic p-type crystal wafer lies between two oxide layers. Note that the vertical P channels isolate the capacitor, the transistor and the resistive elements. The isolation block is readily seen in Fig. A-5. The purpose of the $n+$ layer between the n layer and the p-type crystal wafer is to decrease the series collector resistance.

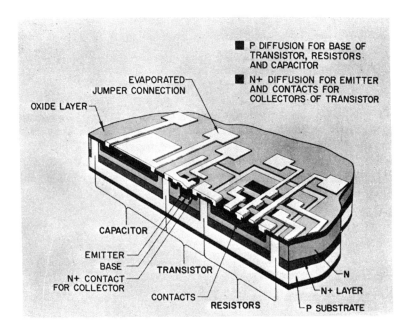

Figure A-6   *Typical Integrated Circuit (Courtesy of RCA Service Co.)*

Fig. A-6 illustrates all circuit elements and the simplicity of circuit connections to the resistor, transistor electrodes and the capacitor.

PACKAGING

Once the integrated circuit has been completed, the resultant wafer must be properly packaged. Two common methods of packaging used in communication systems are the Top Hat or TO type package shown in Fig. A-7(a), or the flat pack shown in Fig. A-7(b). Note that the size of the integrated circuit used is compared with a U. S dime.

APPLICATIONS

In March 1966, RCA was the first manufacturer in the U. S. to introduce integrated circuitry in TV receivers. The single ten-lead device replaced 26 conventional circuit elements in the sound i.f. section of the TV receiver. Shown schematically, the integrated circuit will be represented by a triangle as shown in Fig. A-8.

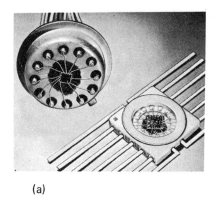

(a)

(b)

Figure A-7   (a) *TO Case and Flat Pack*
(b) *Comparison of TO Case with U. S. Dime*
(*Courtesy of RCA Service Co.*)

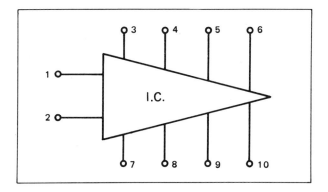

Figure A-8   *Symbol for the IC*

The schematic representation of the elements contained within the IC symbol is shown in Fig. A-9. In this schematic, all of the components enclosed by the dotted line are contained in the symbol in Fig. A-8 for the

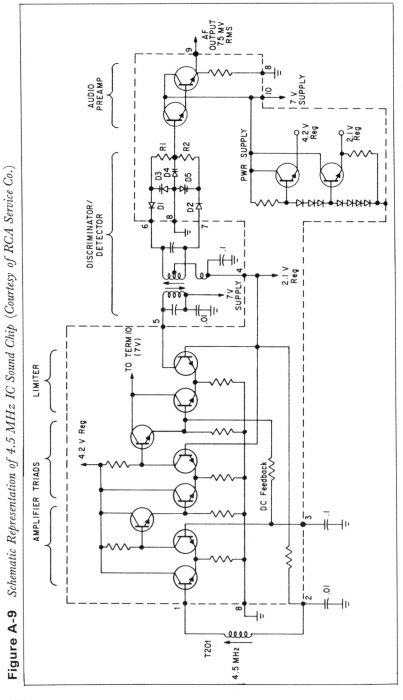

Figure A-9 Schematic Representation of 4.5 MHz IC Sound Chip (Courtesy of RCA Service Co.)

integrated circuit. All other external elements are connected to one of the IC's ten terminal connections.

Note that a 4.5 MHz signal is fed into terminals 1-2 of the IC to drive the sound i.f. and limiter sections of the receiver. This section requires eight transistors with the associated circuit components. Dc coupling is accomplished throughout by use of three transistor combinations using two transistors with emitter coupling and the third transistor as an emitter follower. The limiting process is accomplished in the final emitter-coupled pair. The 4.5 MHz limited output appears at pin 5 of the IC or the input to the discriminator/detector circuit.

Diodes D_1 and D_2 are the key elements of the FM, discriminator circuit. Diodes D_3, D_4 and D_5 do not function as diodes but as capacitors since they are maintained in the reverse biased operating condition. Typical capacity values for the diodes are approximately 7 pF each. The detector network drives a two-stage emitter follower audio preamplifier and provides an output of 75 mV (rms) at output terminal 9.

TROUBLESHOOTING PROCEDURES

The IC is simpler and easier to service than the conventional circuitry in that intermittent connections are either greatly reduced or nonexistent. The problems of aging of components or intermittent operations are completely gone. ICs will either operate properly or fail entirely. Consequently, with proper voltages applied, insert an input signal at the proper terminals and check the output. If the desired output does not appear, replace the IC.

B

SYSTEM
TROUBLESHOOTING

An examination of the complex and intricate nature of modern practical electronic equipment indicates that technicians must have training in electronic systems. Technicians must learn to reason logically since they are required to maintain these electronic systems. Once the technician has absorbed the knowledge of system operation and has the required electronic background, he should be able to troubleshoot and maintain all electronic equipment for the proper state of operation.

It is a rare case when the technician is able to determine immediately the exact electronic component responsible for the system failure. Usually, the technician must find a faulty component by an *isolation procedure*. This involves a method of rapidly localizing the system failure to a single stage or circuit. Consequently, a step-by-step procedure is used.

Step 1: To maintain and repair a piece of electronic equipment, you must determine whether the system is functioning properly. A trouble symptom is indicated by an undesirable change in system performance.

Step 2: Each major section of the system should be tested for required signal output. This procedure will localize the system trouble to a section trouble.

Step 3: Each stage in the defective section should be tested to localize the defective stage in the section failure.

Step 4: The defective stage should be tested to determine the component responsible for the circuit failure.

Troubleshooting methods are not exact in all cases. Tests and observations indicate which stage is most likely the failure unit. Once the failure stage has been determined, appropriate steps should be taken to correct the malfunction. Therefore the localizing procedure is:

1. Localize the system failure to a section.
2. Locate the defective stage within the section.
3. Try a correction.

These steps should be repeated until the equipment is performing properly. Sometimes, the technician may shortcut some of the steps by making a trial correction and hope that he has guessed correctly.

LOCALIZE THE SYSTEM FAILURE

The best help in localizing the trouble to a specific section is the *trouble itself*. In most cases, a malfunction in a particular stage of a complex system is instantly noted by a distinctive change in the system output or effect. Consider a TV set that has no picture, but in which sound and raster are evident. The defect is obviously in the picture section. Therefore, it is important to understand the system equipment function, then tests that eliminate sections are as important as tests that identify a section as malfunctioning.

There are two methods used to localize a malfunction. One method, called *signal tracing*, utilizes the normal signal inherent to the system. This signal is traced until it is lost or *adversely affected* in the defective stage. Signal tracing is performed by connecting a signal generator to the input of the system to be tested.

A suitable output indicating device, usually a voltmeter or an oscilloscope, is then connected to the output of the first stage as shown in Fig. B-1. This process is continued until the indicator device shows abnormal or no output from a stage. This stage is then defective and requires repair.

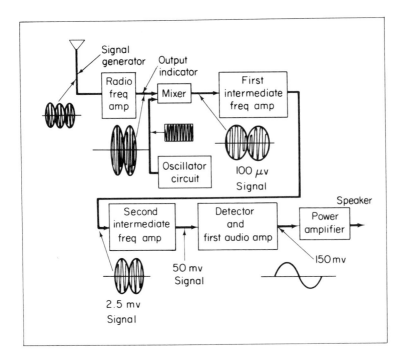

Figure B-1   *Simplified Servicing Block Diagram for Receiver*

A second method uses a signal injection procedure in which the meter or output indicating device is not moved but remains connected to the output. The signal input is injected first to the input of the final stage or power amplifier stage. If the meter indication is satisfactory, the signal is next injected to the stage preceding the final and so on back to the first stage.

As an alternative to using a meter, the output of the system could be heard on the loudspeaker. This method is usually used in *field troubleshooting*.

Once a fault has been localized to a specific stage, voltage and resistance measurements are then performed to pinpoint the defective component.

It should be noted in transistor circuitry, the transistor is usually checked as a last resort. Transistors are quite reliable and since they are generally soldered into the circuit cannot be conveniently checked. Two voltage measurements are used to determine the bias conditions of the stage, namely, V_{ce} and V_{be}. An abnormal collector voltage V_{ce} indicates a faulty circuit element or a change in the collector current flow. If we assume that the transistor itself is performing properly, then any abnormality in conduction is due to a failure of the biasing circuit.

The measurement of V_{be} may present problems since small changes in V_{be} produce large changes in the collector current. Consequently, collector current can be measured indirectly without breaking the circuit by measuring the voltage drop across a resistor in either the emitter or collector circuits. It is then a simple matter of applying Ohm's Law and determine the current flowing through the resistor. For example, in Fig. B-2, a 0.6 volt drop appears across the 500-ohm resistor. The current I_E is equal to:

$$I_E = \frac{0.6}{500} = 1.2 \text{ mA}$$

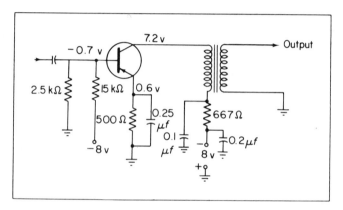

Figure B-2   *Typical Voltages*

The collector current is assumed approximately equal to the emitter current and can be determined from the voltage drop across the 667-ohm resistor. Thus, the voltage drop is $(8 - 7.2) \text{ V} = 0.8 \text{ V}$ assuming the dc resistance of the transformer is negligible. The collector current is:

$$I_C = \frac{0.8}{667} = 1.2 \text{ mA}$$

A change in I_C of more than plus or minus fifty percent ($\pm 50\%$) suggests a failure within the bias system. On most schematics, voltages are given for no signal conditions.

Comparatively high or low collector voltages are indicative of a component failure. For instance, an open emitter resistor will cut off the collector current and set the collector voltage equal to the supply voltage.

Note that if the base divider circuit is defective, then the result is a reduced collector current and collector voltage then tends to rise to the supply voltage. High collector current produces low collector voltage. This condition is produced by a shorted input coupling capacitor or shorted emitter bypass capacitor. Both of these failures will cause an increase in the biasing circuit and produce low collector voltage.

IDENTIFYING TRANSISTOR LEADS

Many case and base wiring systems are in operation at the present time. If possible, manufacturers' data should be considered and consulted for proper lead identification. Some of the popular case and base systems used are shown in Fig. B-3.

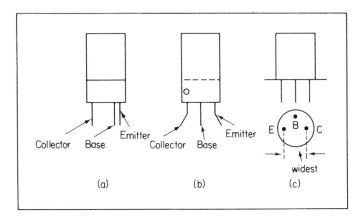

Figure B-3   *Base Connections*

Fig. B-3(a) shows a transistor that is designed for transistor socket usage. Note that the lead standing by itself is the collector; the base lies between the collector and emitter terminals. Fig. B-3(b) shows three equally spaced leads arranged in a line. The collector terminal is identified in this case by a red dot at one end of the case. The base again is the center lead. The triangular shaped arrangement shown in Fig. B-3(c) is quite commonly used. The leads may be identified by holding the semiconductor in the left hand, leads upward and rotating until the distance or base leg of the triangular shape is the largest. Then the emitter terminal is on the left with the base and collector leads identified by counting clockwise from the emitter. This procedure is illustrated in Fig. B-3(c).

Many power transistors use the case style shown in Fig. B-4. The collector terminal is returned to the case itself. Connections to the collector are made by mounting bolts or by a centrally located threaded stud. The base and emitter leads are identified by holding the case so that the two wire leads are to the left of center as shown. Then the emitter is the top lead and the base is the bottom lead.

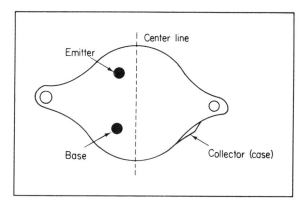

Figure B-4 *Base Diagram for a Power Transistor*

TESTING TRANSISTORS

A transistor that is operated properly, within its current and voltage ratings according to the manufacturer, should have almost unlimited life expectancy. Improperly functioning circuits are usually caused by a failure or malfunction of some circuit component. Consequently, in spite of the fact that the transistor is capable of extremely long life, failures may occur because of:

1. Shorts or open circuits in the biasing section.
2. Temporary power supply overloading.
3. Poor servicing technique.

Thus, it is evident that it may be necessary to evaluate the transistor performance. From the technician's viewpoint, a few simple tests should be adequate to determine the vast majority of transistor trouble possibilities.

The transistor contains two PN junctions or diodes. Damage to the transistor almost invariably is indicated by a malfunctioning of one of the junctions. The transistor failure may be caused by an open or shorted junction or excessive leakage current.

A rapid approximation of the condition of the junctions can readily be made with the ohmmeter section of the multimeter. The forward resistance of each junction should be measured. The method of connecting the ohmmeter terminals to the transistor is shown in Fig. B-5.

Note that the negative terminal of the ohmmeter is always connected to the base for a PNP-type transistor. The forward resistance of both junctions can be checked by touching the positive lead to the emitter and then

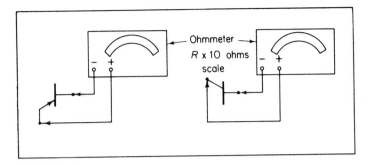

Figure B-5 *Method of Testing Junction Forward Resistance*

to the collector terminals in turn. A normal unit should have a reading below 500 ohms. A high reading indicates an open circuit junction. The forward resistance of the junctions of an NPN-type transistor is checked in the same manner but with the leads reversed; that is, the positive terminal of the ohm-meter is connected to the base and the negative terminal is touched to the emitter and to the collector terminals in turn.

A method of checking short circuits or excessive leakage current requires the reversal of the ohmmeter connections. Also, since the resistance will be higher, the scale should be set at $R \times 10$ kohms as shown in Fig. B-6.

Now the ohmmeter places a reverse bias on each junction in turn. Low and medium power germanium transistors should read about 500 kohms. Silicon transistors give much higher resistance readings. Power transistors have larger junctions and therefore greater leakage currents, with resultant lower resistance readings. Reverse bias resistance readings should be about 50 kohms or greater for power transistors. It should be noted that the actual numerical value in ohms is worthless and should be used as an indication only for PN junction open and short circuits.

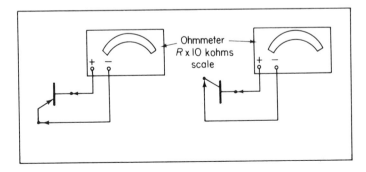

Figure B-6 *Method for Testing Reverse Resistance of Both Junctions*

C

STANDARD COLOR CODING

RESISTORS

The basic system used to identify small size resistors (under two watts) is shown in Table 1. Resistors of this size have either axial leads or radial leads. Three colors are used on each type of resistor to identify its ohmic value. The tolerance marking may vary.

TABLE C-1 *RESISTORS*

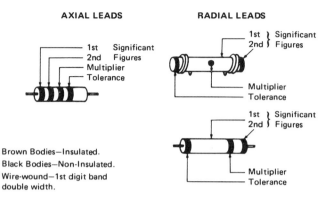

AXIAL LEADS

- 1st — Significant
- 2nd — Figures
- Multiplier
- Tolerance

Brown Bodies—Insulated.
Black Bodies—Non-Insulated.
Wire-wound—1st digit band
double width.

RADIAL LEADS

- 1st ⎱ Significant
- 2nd ⎰ Figures
- Multiplier
- Tolerance

- 1st ⎱ Significant
- 2nd ⎰ Figures
- Multiplier
- Tolerance

COLOR	DIGIT	MULTIPLIER	TOLERANCE
RESISTANCE IN OHMS			
BLACK	0	1	±20%
BROWN	1	10	±1%
RED	2	100	±2%
ORANGE	3	1000	±3%*
YELLOW	4	10,000	GMV*
GREEN	5	100,000	±5% (EIA Alternate)
BLUE	6	1,000,000	±6%*
VIOLET	7	10,000,000	±12½%*
GRAY	8	.01 (EIA Alternate)	±30%*
WHITE	9	.1 (EIA Alternate)	±10% (EIA Alternate)
GOLD		.1 (JAN and EIA Preferred)	±5% (JAN and EIA Pref.)
SILVER		.01 (JAN and EIA Preferred)	±10% (JAN and EIA Pref.)
NO COLOR			±20%

*GMV = guaranteed minimum value, or –0 + 100% tolerance.
±3%, 6%, 12½%, and 30% are ASA 40, 20, 10, and 5 step tolerances.

Applications

Resistance	1st	2nd	Multiplier
68	Blue	Gray	Black
470	Yellow	Violet	Brown
1,500	Brown	Green	Red
22,000	Red	Red	Orange
100,000	Brown	Black	Yellow

CAPACITORS

The basic system used to identify capacitors utilizes the same code (color) as do resistors. Capacitors use additional color bands or dots for other information. Table 2 illustrates the types of ceramic capacitors used in practice.

TABLE C-2

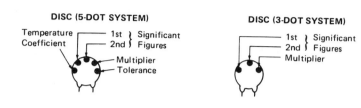

DISC (5-DOT SYSTEM)

Temperature Coefficient — 1st } Significant / 2nd } Figures — Multiplier — Tolerance

DISC (3-DOT SYSTEM)

1st } Significant / 2nd } Figures — Multiplier

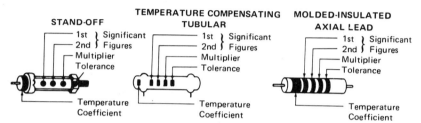

STAND-OFF

1st } Significant / 2nd } Figures — Multiplier — Tolerance — Temperature Coefficient

TEMPERATURE COMPENSATING TUBULAR

1st } Significant / 2nd } Figures — Multiplier — Tolerance — Temperature Coefficient

MOLDED-INSULATED AXIAL LEAD

1st } Significant / 2nd } Figures — Multiplier — Tolerance — Temperature Coefficient

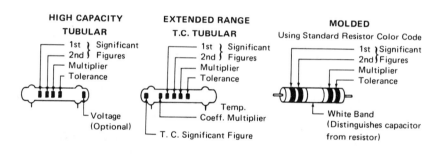

HIGH CAPACITY TUBULAR

1st } Significant / 2nd } Figures — Multiplier — Tolerance — Voltage (Optional)

EXTENDED RANGE T.C. TUBULAR

1st } Significant / 2nd } Figures — Multiplier — Tolerance — Temp. Coeff. Multiplier — T. C. Significant Figure

MOLDED

Using Standard Resistor Color Code

1st } Significant / 2nd } Figures — Multiplier — Tolerance — White Band (Distinguishes capacitor from resistor)

COLOR	DIGIT	MULTI-PLIER	TOLERANCE		TEMPERATURE COEFFICIENT PPM/°C	EXTENDED RANGE TEMP. COEFF.	
			10 μμ FD. or less	Over 10 μμ FD.		Sig. Fig.	Multi-plier
BLACK	0	1	±2.0 μμ FD.	±20%	0 (NPO)	0.0	-1
BROWN	1	10	±0.1 μμ FD.	±1%	-33 (N033)		-10
RED	2	100		±2%	-75 (N075)	1.0	-100
ORANGE	3	1000		±2.5%	-150 (N150)	1.5	-1000
YELLOW	4	10,000			-220 (N220)	2.2	-10,000
GREEN	5		±0.5 μμ FD.	±5%	-330 (N330)	3.3	+1
BLUE	6				-470 (N470)	4.7	+10
VIOLET	7				-750 (N750)	7.5	+100
GRAY	8	.01	±0.25 μμ FD.		+30 (P030)		+1000
WHITE	9	.1	±1.0 μμ FD.	±10%	General Purpose		+10,000
SILVER					Bypass & Coupling		
GOLD					+100 (P100, JAN)		

CAPACITY IN μμ F.

Voltage ratings are standard 500 volts for some manufacturers, 1000 volts for other manufacturers.

TABLE C-3 *APPLICATION OF CAPACITOR COLOR CODE*

Cap. in pF	1st	2nd	Multiplier	Tolerance
47	Yellow	Violet	Black	Brown (1%)
240	Red	Yellow	Brown	Orange (2.5%)
1000	Brown	Black	Red	Green (5%)
4500	Yellow	Green	Red	White (10%)
6800	Blue	Gray	Red	Black (20%)
10,000	Brown	Black	Orange	Black (20%)

TRANSFORMERS

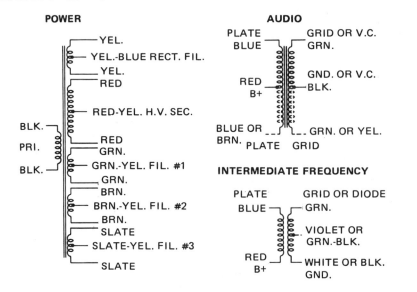

WIRING

COLOR	CIRCUITS
BLACK	GROUNDS
BROWN	FILAMENTS
	HEATERS
RED	B-PLUS
ORANGE	SCREEN GRIDS
YELLOW	CATHODES
GREEN	CONTROL GRIDS
BLUE	PLATES
VIOLET	
GRAY	A.C. LINES
WHITE	OFF-GROUND
	RETURNS

BIBLIOGRAPHY

Chapter 1

1. Shockley, W., *Electrons and Holes in Semiconductors*, D. Van Nostrand, Princeton, N.J., 1950.
2. Shockley, W., *Transistor Electronics—Imperfections in Unipolar and Analog Transistors*, Proc. IRE Vol. 40, p. 1303, 1952.
3. Dekker, A. J., *Solid State Physics*, Prentice-Hall, Inc., Englewood Cliffs, N.J., 1957.
4. DeWitt and Rosoff, *Transistor Electronics*, McGraw-Hill Book Co., N.Y., 1957.
5. Nussbaum, A., *Semiconductor Device Physics*, Prentice-Hall, Inc., Englewood Cliffs, N.J., 1962.
6. Millman and Halkias, *Electronic Devices and Circuits*, McGraw-Hill Book Co., N.Y., 1967.
7. Ryder, J. D., *Electronic Fundamentals and Applications*, 3rd Ed., Prentice-Hall, Inc., Englewood Cliffs, N.J., 1964.

Chapter 2

1. Bardeen and Brattain, W. H., *The Transistor, a Semiconductor Triode*, Physics Review, Vol. 74, p. 230, 1948.
2. Proceedings of the IRE Special Transistor Issue, Vol. 40, 1952.
3. Middlebrook, R. D., *Introduction to Junction Transistor Theory*, J. Wiley and Sons, N.Y., 1957.
4. Searle, C. L. (et. al.), *Elementary Circuit Properties of Transistors*, SEEC, Vol. 3, J. Wiley and Sons, N.Y., 1964.
5. Millman and Halkias, *Electronic Devices and Circuits*, McGraw-Hill Book Co., N.Y., 1967.
6. Ryder, J. D., *Electronic Fundamentals and Applications*, 3rd Ed., Prentice-Hall, Inc., Englewood Cliffs, N.J., 1964.

Chapters 3 and 4

1. Dion, D. F., *Common Emitter Transistor Amplifier*, Proc. IRE, Vol. 46, May, 1958.
2. Giacolletto, L. J., *Study of PNP Alloy Junction Transistors from DC through Medium Frequencies*, RCA Review, Vol. 15, 1954.
3. Levine, I., *High Input Impedance Transistor Circuits*, Electronics, Vol. 33, 1960.
4. Dacey, G. C. and Ross, I. M., *The Field Effect Transistor*, Bell System Tech. Jour., Nov., 1955.
5. Pettit, and Whorter, M. M., *Electronic Amplifier Circuits*, McGraw-Hill Book Co., N.Y., 1961.
6. Phillips, A. B., *Transistor Engineering*, McGraw-Hill Book Co., N.Y., 1962.
7. Shea, R. F. (et. al.), *Principles of Transistor Circuits*, J. Wiley and Sons, N.Y., 1953.
8. Corning, J. J., *Transistor Circuit Analysis and Design*, Prentice-Hall, Inc., Englewood Cliffs, N.J., 1965.
9. Fitchen, F. C., *Transistor Circuit Analysis and Design*, 2nd Ed., D. Van Nostrand, Princeton, N.J., 1966.
10. Seely, S., *Electronic Circuits*, Holt, Rinehart and Winston, N.Y., 1968.
11. Wallmark, J. T., and Johnson, H., *Field Effect Transistors*, Prentice-Hall, Inc., Englewood Cliffs, N.J., 1966.
12. Zeines, B., *Electronic Communications Systems*, Prentice-Hall, Inc., Englewood Cliffs, N.J., 1970.

Chapter 5

1. Henkels, H. W., *Germanium and Silicon Rectifiers*, Proc. IRE, Vol. 46, 1958.
2. Schade, O. H., *Analysis of Rectifier Operation*, Proc. IRE, Vol. 31, 1943.
3. Waidelich, D. L., *Analysis of Full Wave Rectifier and Capacitive Input Filter*, Electronics, Vol. 20, Sept., 1947.
4. Millman, J., and Halkias, C., *Electronic Devices and Circuits*, McGraw-Hill Book Co., N.Y., 1967.
5. Ryder, J. D., *Electronic Fundamentals and Applications*, 3rd Ed., Prentice-Hall Inc., Englewood Cliffs, N.J., 1964.
6. *Silicon Rectifier Handbook*, Motorola, Phoenix, Ariz., 1966.

Chapter 6

1. Armstrong, L. D., and Jenny, D. A., *Behaviour of Germanium Junction Transistors at Elevated Temperatures and Power Transistor Design*, Convention Record, IRE, 1953.
2. Fitchen, F. C., *Transistor Circuit Analysis and Design*, 2nd Ed., D. Van Nostrand, Princeton, N.J., 1966.
3. Hall, R. N., *Power Rectifiers and Transistors*, Proc. IRE, Vol. 40, 1952.
4. Corning, J. J., *Transistor Circuit Analysis and Design*, Prentice-Hall Inc., Englewood Cliffs, N.J., 1965.

5. Jones and Hilbourne, *Transistor A. F. Amplifiers*, New York Philosophical Library, 1957.
6. Shea, R. F., *Transistor Audio Amplifiers*, J. Wiley and Sons, N.Y., 1955.
7. Seeley, S., *Electronic Circuits*, Holt, Rinehart and Winston, N.Y., 1968.

Chapters 7 and 8

1. Blecher, F. H., *Design Principles for Single Loop Transistor Feedback Amplifiers*, IRE Transactions Circuit Theory, CT-4, Sept., 1954.
2. Gibbons, J. F., *Semiconductor Electronics*, McGraw-Hill Book Co., N.Y., 1966.
3. Hakim, S. S., *Junction Transistor Analysis*, J. Wiley and Sons, N.Y., 1962.
4. Hollman, H. E., *Transistor Oscillators*, TeleTech, 12 Oct., 1953.
5. Hunter, L. P., *Handbook of Semicondictor Electronics*, McGraw-Hill Book Co., N.Y., 1956.
6. Joyce, M. V., and Clarke, K. K., *Transistor Circuit Analysis*, Addison Wesley, Reading, Mass., 1961.
7. Page, D. F., *A Design Basis for Junction Transistor Oscillator Circuits*, Proc. IRE, Vol. 46, 1958.
8. Pullen, K., *Handbook of Transistor Circuit Design*, Prentice-Hall, Inc., Englewood Cliffs, N.J., 1961.
9. Shea, R. F., *Transistor Circuit Engineering*, J. Wiley and Sons, N.Y., 1967.
10. Seeley, S., *Electronic Circuits*, Holt, Rinehart and Winston, N.Y., 1968.

Chapters 9 and 10

1. Bedrosian, E., *The Analytic Signal Representation of Modulated Waveforms*, Proc. IRE, Oct., 1962.
2. Bennet, W. R., and Davey, J. R., *Data Transmission*, McGraw-Hill Book Co., N.Y., 1965.
3. Black, H. S., *Modulation Theory*, D. Van Nostrand, Princeton, N.J., 1953.
4. Downing, J. J., *Modulation Systems and Noise*, Prentice-Hall, Inc., Englewood Cliffs, N.J., 1964.
5. Hancock, J., *Principles of Communication Theory*, McGraw-Hill Book Co., N.Y., 1961.
6. Lathi, B. P., *Signals, Systems and Communication*, J. Wiley and Sons, N.Y., 1965.
7. Loughlin, B. D., *Theory of Amplitude Modulation Reaction in the Ratio Detector*, Proc. IRE, March, 1952.
8. Reza, F. M., *An Introduction to Information Theory*, McGraw-Hill Book Co., N.Y., 1961.
9. Schwartz, M., *Information Transmission, Modulation and Noise*, 2nd Ed., McGraw-Hill Book Co., N.Y., 1970.
10. Taub, H., and Schilling, D. L., *Principles of Communication Systems*, McGraw-Hill Book Co., N.Y., 1971.
11. Wozencraft, J. M., and Jacobs, I. M., *Principles of Communication Engineering*, J. Wiley and Sons, N.Y., 1965.

ANSWERS

Chapter 1

1. $I = 15.54$ mA
3. $E = 168$ mV
5. $T = 62.8°C$
7. $I_0 = 3.64\,\mu A$
9. (a) $T_F = 36.7°C$
   (b) $I = 514\,\mu A$
   (c) $r = 48.2\,\Omega$
11. $E = 243$ mV
13. $T = -61°C$

Chapter 2

1. $v_{ce} = 20$ V
   $I_c = 5$ mA
3. $I_c = 2.34$ mA
   $I_b = 36.53\,\mu A$
   $I_E = 2.376$ mA
5. (a) $V_{cq} = 7.8$ v
   $I_{cq} = 4.2$ mA
   (b) $v_{cq} = 8.9$ V
   $I_{cq} = 5.5$ mA
7. (a) $R = 134.5$ kΩ
   (b) Load line
   (c) $S = 36.7$
9. $S = 10$
11. (a) $V_{cq} = 7$ V
   $I_{cq} = 4.5$ mA
   (b) $V_{cq} = 1.8$ V
   $I_{cq} = 5.5$ mA
13. $S = 7.26$
15. $S = 25.23$
17. $S = 10.075$
19. $R_E = 625\,\Omega$
   $R_1 = 95.6$ kΩ
   $R_2 = 10.05$ kΩ

Chapter 3

1. (a) $z_{12} = 3-j6 = z_{21}$
   $z_{11} = z_{22} = 6-j_{12}$
   (b) $z_{12} = z_{21} = 4.8 + j\,4.6$ ohms.
   $z_{11} = 5.7 + j\,5.3$ ohms.
   $z_{22} = 4.8 + j\,2.6$ ohms.
3. (a) $y_{11} = y_{22} = (44.5 + j\,89)10^{-3}$ mhos.
   $y_{21} = y_{12} = (49 - j\,8.9)10^{-3}$ mhos.
   (b) $y_{11} = .29 + j\,.332$ mhos.
   $y_{12} = -(.19 + j\,.416)$ mhos.
   $y_{21} = -(.19 + j\,.416)$ mhos.
   $y_{22} = .229 + j\,.487$ mhos.
5. $A_i = 44.5$
   $R_i = 1224\,\Omega$
   $A_v = -121$
   $P.G. = 5380$
   $R_0 = 38.2$ kΩ
7. $y_{11} = .148\,\underline{)10.2°}$ mhos.
   $y_{12} = .103\,\underline{)151°}$ mhos.
   $y_{21} = .103\,\underline{)173.6°}$ mhos.
   $y_{22} = .143\,\underline{)-9.7°}$ mhos.
9. $h_{11} = 5.38\,\underline{)60.3°}$ ohms.
   $h_{12} = .473\,\underline{)-45°}$
   $h_{21} = .85\,\underline{)101.3°}$
   $h_{22} = .236\,\underline{)-450°}$ mhos.
11. $h_{11} = 2.55\,\underline{)229.5°}$ ohms.
   $h_{12} = 1.12\,\underline{)60°}$
   $h_{21} = 2.05\underline{)136°}$
   $h_{22} = .513\,\underline{)-15.75°}$ mhos.

Chapter 4

1. (a) $A_{i_{mid}} = 13.33$
   (b) $f_l = 72.64$ Hz
   (c) $f_h = 297$ kHz.
3. (a) $A_{i_{mid_T}} = 177.7$

366

(b) $f_{l_T} = 113$ Hz.

(c) $f_{h_T} = 191$ kHz.

5. Proof.
7. (a) $V_0 = .91 V_2$
   (b) $V_0 = -.7 V_1$
9. $V_{0_1} = .263 V_i$
   $V_{0_2} = -9.21 V_i$
11. (a) $A_{v_{mid}} = -37.5$
    (b) $f_h = 84.9$ kHz.
    (c) $C_c = .035 \mu F$
13. (a) $A_{v_{mid}} = -28.57$
    (b) $f_l = 13.64$ Hz.
    (c) $f_h = 37.14$ kHz.
    (d) $A_v = -23.4$
15. (a) $R_n = 1600 \Omega$
    $C_n = 1.99$ pF
    (b) $A_{v_{max}} = 67.5$
    (c) $\Delta f = 1.31$ MHz.
    (d) $p = .9$
17. (a) $L = 2.88 \mu H$
    (b) $A_{v_{max}} = -79.5$
    (c) $p = .928$
19. (a) $A_{v_{max}} = -10.75$
    (b) $\Delta f = 15.45$ MHz
    (c) $p = .613$
21. (a) $\Delta f = 1.565$ MHz
    (b) $L = .944 \mu H$
    $C = 29.8$ pF
    (c) $p = .86$
23. (a) $\Delta f = 2.69$ MHz
    (b) $A_{v_{max}} = -213.7$
    (c) $p = .561$
25. (a) $\Delta f = 1.99$ MHz
    (b) $p = .92$

Chapter 5

1. (a) $I_{dc} = 811$ mA
   (b) $V_{rms_{ac}} = 18.65$ V
   (c) $Z = 38.5\%$
3. (a) $I_{dc} = 902$ mA
   (b) $V_{rms_{ac}} = 19.5$ V
   (c) $Z = 73\%$
5. (a) $I_{dc} = 131$ mA
   (b) $V_{rms_{ac}} = 158$ V
   (c) $Z = 73.8\%$

7. (a) $I_{dc} = 144$ mA
   (b) $V_{rms_{ac}} = 36.8$ V
9. (a) $V_{dc} = 555.2$ V
   (b) $V_{rms_{ac}} = 6.05$ V
11. (a) $V_{dc} = 315$ V
    (b) $I_{dc} = 105$ mA
    (c) $V_{rms_{ac}} = 334$ mV
13. $V_{dc} = 360$ V
    $V_{rms_{ac}} = 298.5$ mV
15. $V_{dc} = 360$ V
    $V_{rms_{ac}} = 374$ mV
17. $V_{dc} = 515.6$ V
    $V_{rms_{ac}} = 35$ mV
19. $V_{dc} = 515.6$ V
    $V_{rms_{ac}} = 106$ mV
21. $V_{dc} = 360$ V
    $V_{rms_{ac}} = 206 \mu V$
23. $V_{dc} = 360$ V
    $V_{rms_{ac}} = 11.24$ V
25. $V_{dc} = 555.7$ V
    $V_{rms_{ac}} = .876$ V
27. $V_{dc} = 360$ V
    $V_{rms_{ac}} = 16.8$ V

Chapter 6

1. $R_{ac} = 3600 \Omega$
3. (a) $P_0 = 4$W
   (b) $V_{P_{rms}} = 5.66$ V
   (c) $V_{S_{rms}} = 113.2$ V
5. $D_2 = 3.84\%$
7. $D_2 = 5.78\%$
9. $D_2 = 10.4\%$
   $D_3 = -4.167\%$
   $D_4 = 6.25\%$
   $D_t = 12.9\%$

Chapter 7

1. (a) $A_{v_r} = -9.09$
   (b) $f_l = 18.18$ Hz
   (c) $f_h = 330$ kHz
3. (a) $A_v = -100$
   (b) $f_l = 200$ Hz
   $f_h \doteq 80$ kHz

5. (a) $A_{v_r} = -2.73$
   (b) $R_{in} = 73.5\ K\Omega$
7. (a) $A_{v_r} = -9.09$
   (b) $r_{d_r} = 551\ k\Omega$
9. (a) $A_{v_r} = -9$
   (b) $r_{d_r} = 141.5\ k\Omega$
11. (a) $A_{i_r} = -3$
    (b) $R_F = 20\ K$
13. (a) $A_{v_{mid}} = -30$
    (b) $\mu_r = 5.4$
    $r_{d_r} = 4.1\ k\Omega$
15. (a) $B_1 = .0644$
    $B_2 = .222$
    (b) $A_{v_r} = -3.17$
    (c) $\mu_r = 4.3$
    $r_{d_r} = 11.3\ k\Omega$

Chapter 8

1. $w_{osc} = \dfrac{1}{\sqrt{(L_1 + L_2)\,C}}$

   $h_{fe_{req}} = L_2/L_1$
3. $w_{osc} = 11.7\ \text{Mrad/sec.}$
   $\mu_{req} = 2.04$
5. $w_{osc} = 7.74\ \text{Mrad/sec.}$
   $\mu_{req} = .4975$

7. $w_{osc} = \dfrac{1}{\sqrt{(L_1 + L_2)C_3 + L_1 C_1}}$

   $\mu_{req} = \dfrac{L_2}{L_1}$
9. $w_{osc} = \sqrt{6}R/L$
   $A_v = -29$

Chapter 9

1. $P_c = 10kW$
   $P_{LSB} = P_{USB} = 100\ W.$
3. (a)

Peak Amp.	freq in 14Hz.
50	1
18.75	$\begin{cases}1.001 \\ .999\end{cases}$
11.25	$\begin{cases}1.002 \\ .998\end{cases}$

(b) $P_c = 25W.$
$P_{SB_1} = 7.07\ W.$
$P_{SB_2} = 2.54\ W.$

(c) Graph.
(d) $m_c = .875$
5. $L = 3.18\ \mu H$
   $C = 79.5\ pF$
7. $C_1 = 954\ pF$
   $C_2 = 5160\ pF$
   $L = 30 = H.$
9. Proof.
11. Proof.
13. $de_a = 1.95\ V.$

Chapter 10

1.

Amp. in μA	freq.
29	0
20	$w \pm p$
4	$2p$

3.

Amp. in μA	freq.
836	0
240	$2w \pm p$
836	$2w$
36	$2p$
480	p
18	$2(w \pm p)$

5.

Amp. in μA	freq.
116.5	0
112.5	$2w$
100	p
12.5	$2p$
50	$2w \pm p$
6.25	$2(w \pm p)$

7. From Graph
   (a) $Z_d = 84\%$
   (b) $R_{eff} = 125\ k\Omega$
   (c) $A.V.C. = 21\ V$
9. From Graph
   (a) $Z_{d_{dc}} = 84\%$
   $Z_{d_{ac}} = 82\%$
   (b) $R_{eff} = 71.04\ k\Omega$
   (c) $A.V.C. = 21\ V$
11. (a) $V_0 = 17.14\ V$
    (b) $V_0 = 10.75\ V$
    (c) $V_0 = 8.7\ V$
    (d) $V_0 = 4.5\ V$
    (e) $V_0 = .64\ V$

INDEX

LIBRARY
FLORISSANT VALLEY COMMUNITY COLLEGE
ST. LOUIS, MO.

INVENTORY 1983